T0245337

Chimeras and Consciousness

Chimeras and Consciousness

Evolution of the Sensory Self

edited by Lynn Margulis, Celeste A. Asikainen, and
Wolfgang E. Krumbein

The MIT Press
Cambridge, Massachusetts
London, England

Set in Sabon by Toppan Best-set Premedia Limited.

Library of Congress Cataloging-in-Publication Data

Chimeras and consciousness : evolution of the sensory self / edited by Lynn Margulis, Celeste A. Asikainen, and Wolfgang E. Krumbein.
    p. cm.
Includes bibliographical references and index.
ISBN 978-0-262-01539-4 (hardcover : alk. paper) — ISBN 978-0-262-51583-2 (pbk. : alk. paper)
1. Consciousness. 2. Brain—Evolution. 3. Self-consciousness (Awareness)
I. Margulis, Lynn, 1938– II. Asikainen, Celeste A. III. Krumbein, Wolfgang E.
QP411.C55   2011
612.8'2—dc22
                                                          2010032242

The MIT Press is pleased to keep this title available in print by manufacturing single copies, on demand, via digital printing technology.

Contents

Foreword by John B. Cobb Jr.   ix
Preface   xiii
Acknowledgments   xvii

Introduction: Life's Sensibilities   1

I Selves

1 **Valuable Viruses   17**
  Frank P. Ryan

2 **More Like a Waterfall   23**
  William Day

3 **Alarmones   35**
  Antonio Lazcano, Arturo Becerra, and Luis Delaye

4 **Early Sensibilities   45**
  Kenneth H. Nealson

II Groups

5 **Smart Bacteria   55**
  Eshel Ben-Jacob, Yoash Shapira, and Alfred I. Tauber

6 **Ancient Architects   63**
  Wolfgang E. Krumbein and Celeste A. Asikainen

7 **Others   71**
  Laurie Lassiter

8 **Nested Communities   91**
  James MacAllister

III Earth

9 Cosmic Rhythms of Life   109
Bruce Scofield

10 Life's Tectonics   123
Paul D. Lowman Jr. and Nathan Currier

11 Evolutionary Illumination   129
Peter Warshall

IV Chimeras

12 Symbiogenesis in Russia   153
Victor Fet

13 From Movement to Sensation   159
John L. Hall and Lynn Margulis

14 Packaging DNA   167
Andrew Maniotis

15 Lemurs and Split Chromosomes   173
Robin Kolnicki

16 Interspecies Hybrids   183
Sonya E. Vickers and Donald I. Williamson

17 Origins of the Immune System   199
Margaret J. McFall-Ngai

18 Medical Symbiotics   207
Jessica Hope Whiteside and Dorion Sagan

V Consciousness

19 Animal Consciousness   221
Gerhard Roth

20 Brains and Symbols   233
John Skoyles

21 Thermodynamics and Thought   241
Dorion Sagan

22 "I Know Who You Are, I Know Where You Live"   251
Judith Masters

23 Cultural Networks   259
Luis Rico

Bibliography   267
Appendix A: Major Groups of Living Organisms   281
Appendix B: The International Geological Time Scale   289
Glossary   291
About the Authors   309
Index   315

Foreword

I am honored to write the foreword for this book. This invitation indicates that at least at some points the exclusion of humanistic concerns from the sciences is breaking down. Here even a theologian has been invited. I look forward to the day when it will be widely recognized both that theology is too important to be left to professional theologians and that science is too important to be left to professional scientists. We need an inclusive vision that makes sense of our experience of the world and of all that science has taught us about it.

I am a philosophical theologian. Like many humanists, I have an understanding of my world that is extensively informed by science; so I have a lay understanding of several fields. One field in which I have been particularly interested is the history of life on the planet, usually called evolutionary biology. I collaborated with an ecologist, Charles Birch, on a book titled *The Liberation of Life*. More recently I edited the contributions that resulted from a small conference that took place at our graduate school of theology (and Whiteheadian philosophy) in Claremont, California, for publication in a book titled *Back to Darwin: A Richer Account of Evolution*. The contributors included biologists (of whom Lynn Margulis was one), other scientists, philosophers, and theologians.

I am a strong supporter of evolution but am quite critical of the way it is usually presented by what we outsiders call "neo-Darwinism." I am opposed to reductionism and determinism, and I find the dominant presentation of evolution to be characterized by both. It is my judgment that this commitment of so much of science derives from the seventeenth-century metaphysics with which modern science grew up rather than from the actual modern observations and data. This metaphysics has been outgrown in advanced regions of physics, but it has maintained its hold on mainstream biology.

When the evolution of species, including the human species, was demonstrated by Darwin, scientists did not change their view of nature as matter in motion but rather simply extended it to include human beings. The implication is that the ideas presented by scientists are to be explained by reducing them to matter in motion rather than attending to their meaning and the detailed evidence in their favor. This *reductio ad absurdum* suggests something I do not believe. We are not zombies. Neither we nor other animals, plants, or microorganisms have evolved from automata. If we take account of animal (including human) experience and activity, this reductionism can be overcome. More of the actual evidence garnered from many scientific fields can be coherently explained. We can study the evolution of what in this book are called "selves" as well as the evolution of bodies.

Both "symbiogenesis" and "Gaia" refer to the activity of living things as causally explanatory of what happens. Hence I am prejudiced in their favor. Of course, I have tried to attend to arguments against them, but these have not seemed strong. I believe that their systematic omission from typical explanations of evolution reflects the metaphysical prejudice of which I have spoken rather than lack of evidence.

All of this is to say that I approach this book with enthusiasm. I know of no other book like it. The closest with which I happen to be familiar is *Evolution in Four Dimensions* by Eva Jablonka and Marian J. Lamb. It is written as a single integrated argument against genetic determinism rather than a collection of papers. But the scientific data provided in it deal with only a small fraction of the subject matter presented here.

The authors of these chapters are quite diverse, but in general the book presents living thing as selves that act rather than as machines manipulated by genes. Genes certainly play an important role, but the writers are open to a much wider range of forces operating with living things and therefore in their evolution. The extensive attention to the early stages of evolution places what zoologists discuss in illuminating perspective.

Not being a scientist, I cannot speak with authority on the subject of the scientific accuracy of all that is said. However, I have very little doubt that the authors are all responsible scientists. They present evidence without forcing it to fit with the scientific metaphysics that plays so large a role in modern biology. They do not go far to draw the conclusions that are of special philosophical interest, but that is all to the good. Nowhere do I feel that data are being pressed into the service of predetermined beliefs. However, the overall impact shows the evolutionary process to be vastly more complex than is now typically assumed.

I wish that I had been aware of more of this material when I worked on *Back to Darwin*. Based on the contributions of the participants, I tried in that book to show the importance of animal activity, much of it purposeful, in shaping the course of evolutionary events. It seemed important to me that some biologists now recognized that the natural selection of particular genetic mutations often followed on changed behavior. However, at the time I was ignorant of most of the material presented in this volume. For example, I knew nothing about the evidence for interspecies hybrids presented in chapter 16. Clearly the theory proposed is still not "proved." But the chapter persuades me that it is the most plausible explanation of the phenomena.

I do not see how anyone could read this book and continue to be satisfied with the simple neo-Darwinist account. As one who thinks that account has done great harm, partly by evoking equally unacceptable reactions, I consider this to be an extremely important collection of papers that could change the nature of the currently unhealthy and unhelpful arguments about evolution.

Obviously, the book has its weaknesses. Many of the chapters are fully accessible only to specialists. Someday I hope there will be writings on evolution for the general public that explain the ideas presented in this book in a widely accessible way. But for now, we need this book, because more popular ones can be easily marginalized as unscientific, not only by scientists but also by general readers. Popular books should reveal new formulations of evolutionary theory among scientists.

The introduction to this book and the abstracts that introduce the individual chapters makes the book's contents accessible to a wider public and clarify the implications. Because of this editorial work, the book is far more than a collection of generally congenial papers. It is a rich introduction to a vast field of research still little known to the general public and insufficiently appreciated by mainstream scientists.

John B. Cobb Jr.
The Center for Process Studies Research Center
Claremont School of Theology
Claremont Graduate University
Claremont, California

Preface

This book, which traces back alleys and main pathways from the earliest life forms to cogitating human beings, stems mainly from an international conference held in 2004 at the Rockefeller Foundation's beautiful Bellagio Center. I believe the subject matter will fascinate those genuinely interested in what is scientifically known about the origin and the evolution of life's consciousness

Although the fields represented here (psychology, microbiology, ecology, history of science, primatology, virology, and geology, among others) are undoubtedly interrelated, *Chimeras and Consciousness*, like science itself, cannot pretend to present a complete or exhaustive picture. Science proceeds by detailed exploration of highly specific aspects of the world. As any scientist explores his discipline, he (or, of course, she) tends to lose sight of the big picture. The short abstracts that introduce the chapters attempt, without simplification and overstatement of the science, to help orient the reader to the chapters' messages.

This book, unlike popular science "trade books," is highly concentrated. It is more like whiskey than beer. It is "the real thing": the chapters are written by real scientists and other scholars without help from journalists, professional educators, or other intermediaries. The usual intermediaries invariably simplify; with the best of intentions, they try to explain ideas that they do not understand. In my experience, it takes so much time to learn firsthand about anything that journalists at all levels of sophistication mostly have to argue from authority. None of the authors are journalists or educators.

Taken together, the chapters resemble a castle glimpsed through a fog in a foreign land. The weather clears. The scene begins to cohere into an awesome structure—reality. Consciousness has been emerging over vast stretches of time. The simpler behaviors and interrelationships of connected living entities have given rise to consciousness as known in humanity only very recently. Nearly four thousand million years of sensitive life everywhere on planet Earth preceded us. Our tropical African talking species appeared less than a million years ago.

Like the symbiotic composite organisms from which we evolved, the chapters treat subjects that must be merged and integrated to form a more complete understanding. Alas, scientific disciplines and ignorance tend to preclude the integration that all of us would like to see. Fields of study—"subjects"—are not present in nature. Rather, nature shows a cavalier disregard for academic disciplines. Nevertheless, this book attempts, fleetingly and imperfectly, to begin to assemble a synthesis of knowledge of the connected world of conscious, sentient, beyond-human life.

The chapters that originated in academic papers presented at an obscure scientific meeting show a diversity of approaches. Together they lead to an emerging understanding of specific aspects of life's interactions in the biosphere over eons.

We should be suspicious of journalistic complete explanations and comprehensive stories, despite their allure. The former suggest too-quick connections; the latter encourages the all-too-human addiction to narrative that prematurely knows the forest without examination of its trees.

I have no illusions that scientific knowledge of reality, despite the know-it-all attitude of our species in general and scientists in particular, is ever really encyclopedic. What we do know is that painstakingly derived measurement and observation often lead us to correct our biases. As an honest scientist, I know I cannot weave a single, seamless narrative from these fascinating threads, despite my wish to do so. Science is not religion, which tends to proffer quick answers to huge questions and then spend time justifying them. Science is the meandering of observation, tentative theories, and bold hypotheses that often prove incorrect. This is its human weakness and its intellectual strength.

Nonetheless, if we were to weave a tapestry of the threads of *Chimeras and Consciousness*, it might be make a flying carpet or a great interstellar sailing ship. I invite you to venture out. Just for now, please overcome your reserve and inhibition. But bring along your sharpest critical faculties. Share with me, just for now, a symbiotic view of life and a nature-embedded perspective of mind. Let us agree that perhaps the three greatest real mysteries of nature are existence, life, and mind. This book leaves the question of existence where it belongs, in the capable hands of theologians and physicists, but it delves deeply into life and mind.

Each of us has proprioceptors—nerves that tell us about our current selves, where we are, if our toes are cold, if our throats are dry, or if we have been deprived of sleep. Our feeling of what it's like to be alive from the inside is certainly a valid scientific sensation. But to what extent is self-awareness shared? How does living matter relate to evolving mind? What happens when we focus attentively on the chemistry and the physiology of nature's details? What do we learn?

In this book we gain valid insight into many aspects of our short lives: our senses of smell and taste, touch and balance, sensing of water and light, and the ancient histories of these senses. We learn about our relations to our home in the Earth-Moon system and its rotation around the Sun. We are surprised that social and sexual phenomena in communities thought to be so quintessentially human have immensely long histories in other forms of life. Often, appropriate study of family members, children and adults who grew up together, leads to predictions of behaviors and fates of the adults in the subsequent generation. Many "facts of life" here are at odds with prejudices and assumptions of our self-centered species in our dominant culture. Strong evidence exists for ideas about which we are unaware or only dimly conscious and which we tend to deny. We consider geology and biology unrelated fields of study, but in fact they are intimately related. We think we tell the truth to our closest kin, but in fact we routinely lie more to them than we lie to total strangers. We see viruses and bacteria as enemies to be conquered and killed, but they are parts of our own ancestry, indeed needed for our own bodies to survive. When I consider the vastness of the topics and the authenticity of the scholar-authors, I find *Chimeras and Consciousness* fascinating, and indeed remarkable for its brevity and accuracy.

Lynn Margulis
Balliol College, Oxford
University of Massachusetts, Amherst

Acknowledgments

We are immensely grateful to those who provided financial support for our Bellagio meeting and for the preparation of this book inspired by it: the Rockefeller Foundation in New York and in Bellagio, and the staff of the Bellagio Study Center, the Deutsche Forschungsgemeinschaft, the Alexander von Humboldt-Stiftung, the Hanse-Wissenschaftskolleg, and the Tauber Fund. At the University of Massachusetts at Amherst, we acknowledge the former deans of the College of Natural Science and Mathematics and Steven Goodwin, dean of the newly organized College of Natural Sciences. We are particularly thankful to Provost Charlena Seymour, formerly dean of the graduate school (now at Simmons College), and to Michael Williams, formerly chairman of our superb Geosciences Department. LM is especially grateful for Balliol College's resident Eastman Professorship privilege in 2008–09.

Colleagues and friends who contributed information and/or direct aid in the preparation of the book include Ian Baldwin, Stephen Bell, Betsy Blunt, Baruch Blumberg, Martin Brasier, Lois Brynes, Karl Campbell, E. Canale-Parola, Emily Case, Michael Chapman, Kendra Clark, Bruce Clarke, Rita Colwell, James di Properzio, Michael Dolan, Galina Dubinina, Paul Evans, Sean Faulkner, Alder Fuller, Andrew Graham, Stephan Harding, Ricardo Guerrero, Judith Hooper, Ching Kung, Christie Lyons, Daniel Guerrero Miracle, Jennifer Margulis di Properzio, Denis Noble, Brian Ogilvie, Jeremy Sagan, Alex Salheny, Jan Sapp, Melishia Santiago, Theodore Sargent, Dennis Searcy, Richard Teresi, William Irwin Thompson, Neil Todd, Dennis Steiner, Alfred Tauber, James Walker, Sarah Whatmore, Andrew Wier, Richard Wilkie, Kathy Willis, and Stewart Wilson. Judith Herrick Beard and Erin Idehenre were fine editor-typists. Dianne Bilyak, our excellent developmental editor, dealt with the contributors of prose, illustrations, references, and other information. We thank Joni Pradad, Brianne Goodspeed, Jonathan

Teller-Ellsberg, Emily Foote, and Margo-the-Magnificent at Chelsea Green Publishing Company. We acknowledge Idalia Rodriguez and Laurie Godfrey, both of the University of Massachusetts at Amherst, for their knowledge of primatology and their generous aid in the preparation of chapter 15. We thank Robert Sternberg and his magnificent documentary film *Hopeful Monsters* for help with the content and the form of chapter 16.

Eileen Crist, Harold Morowitz, Bruce Rinker, the late Stephen Schneider, and John Cobb Jr. helped us land the final manuscript where it belonged: in the hands of our capable, familiar science-environmental editor at the MIT Press. Clay Morgan and his remarkably literate colleagues Laura Callen and Paul Bethge continue to help us complete our "consciousness project," which has been in the works since 2003. Our Environmental Evolution course, begun in 1972 at Boston University, continued in 1989 at the University of Massachusetts at Amherst and will be taught in 2011–12. Since we developed its textbook in its first and second editions (L. Margulis and L. Olendzenski, eds., *Environmental Evolution: The Effect of the Origin and Evolution of Life on Planet Earth*, MIT Press, 1992; L. Margulis, C. Matthews, and A. Haselton, eds., *Environmental Evolution: The Effect of the Origin and Evolution of Life on Planet Earth*, second edition, MIT Press, 2000) and published our earliest Bellagio meeting proceedings (L. Margulis and R. Fester, eds., *Symbiosis as a Source of Evolutionary Innovation: Speciation and Morphogenesis*, MIT Press, 1991), it has always been a pleasure to work with the MIT Press.

Introduction: Life's Sensibilities

Life on Earth is composed of myriad interacting elements, which sense one another and the environment. We humans represent only a tiny part of this colossus. The flying birds, the polarized-light-oriented bees, the bacteria that biomineralize, the gene-trading and fossil-fuel-trading humans, the recycling fungi, the cooling rainforest ecosystem, the oxygen-excreting cyanobacteria, the plankton-raining oceans—that is, geography-changing, energy-transforming, and consciousness-raising activities of life—continue to expand. Life continues to evolve as it has done for more than 3,000 million years. Life differs strikingly from inanimate matter in its possession of sensation, its complex behavior, and its awareness of self. Whether as compact individuals or as widely dispersed slime mold, life forms interact incessantly with their local environment.

Living beings, in their basic material transformations as they cycle energy and matter, organize, and orient to their surroundings, resemble non-living naturally occurring complex systems. Yet we don't feel like a "non-living complex system" from the inside. We feel alive. We are aware, conscious, and often even attentive as we share our cognitive powers and sensibilities with one another and with many kinds of non-human life. The physico-chemical basis of our inner feelings of sensitivity and awareness have been relatively neglected by science. Most professional scientists write, whether about energy flow or psychological mind states, as if their particular field harbored one objective, certain universal truth just waiting to be discovered.

The earliest and most elementary "self" is the prokaryotic (bacterial) cell shown at left in figure I.1. All cells show all the properties of life: they have "identity." All have an inside (the self) and an outside (the environment), defined by their boundary membranes made by the cell itself. Membranes are made of fatty materials (lipids) in which specific proteins are embedded. The entry and exit into and out of the cell of

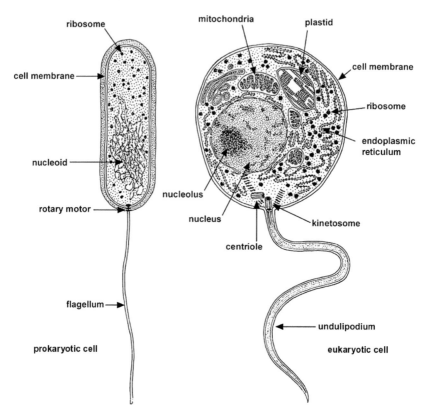

Figure I.1
Cells, minimal autopoetic units of life.

small charged salt ions—either positive ones, such as sodium (Na+), potassium (K^+), magnesium (Mg^{2+}), or calcium (Ca^{2+}), or negative ones, such as chloride (Cl^-), sulfate (SO_4^{2-}), or phosphate (PO_4^{3-})—help maintain the self. Cells are selves; they are systems, and thus each part of the system must always be in place and active to maintain the self. Bacterial cells are sensitive and respond to the major environmental stimuli. These stimuli include water with and without dissolved salt ions. Bacteria respond to gravity, position, hot and cold, and thirst. They respond to sound as movement (they are mechanosensitive) and to taste and smell (they are chemosensitive), and they react to light (they are photosensitive). Living and growing bacterial cells absorb appropriate energy, incorporate carbon and energy, and release particular wastes. Most grow to make larger versions of themselves. Most eventually reproduce or die trying. Sensitivity, awareness, and consciousness correlate with living behavior, which evolutionarily began with bacteria.

The bacteria that we denigrate as "germs" and rally to destroy sense salts and sugars, the slightest wind, the gentlest raindrop, and the incessant downward pull of gravity. "Prototaxis," according to the biologist I. E. Wallin (1927), is the "innate tendency of one organism or cell to react in a definite manner to another organism or cell." Prototaxis evolved in bacteria on an Earth with shorter days and nights, a more influential (because closer and "bigger") Moon, and air depleted in molecular oxygen. This is the environment in which awareness first seems to have flickered into being. Where and how did it do so? How did "awareness" evolve among the interacting cells of life and generate you, who have now begun to read a book on evolution?

We focus here on connected themes: organisms interacting and consciousness arising. What is the source of evolutionary novelty? Contrary to the insistence of neo-Darwinist animal scientists, the accumulation of random mutation in genes is neither the only source of evolutionary novelty nor the major source (Margulis and Sagan 2002). More diverse and original than animals are the bacteria and protoctists of the microscopic world. Inside this world, at our feet beyond and inside us, it is easy to see that life forms of many different shapes and sizes interact, merge, and complexify. Working together to gain access to energy and material for sustenance and growth, organisms unite to form new, larger organisms. Some go on to form societies and communities that fractally complexify on a larger scale. Together they accomplish what would be impossible for them to do on their own. Over the course of evolution, living beings sensed their immediate environments and one another. Cell awareness was here from the time of the origin of earliest life. Awareness and sensation continued to evolve into full-blown consciousness as we humans experience and verbalize it.

One of the great myths of evolution theory in its "modern form of neo-Darwinism" is that traits and behaviors of organisms can be well understood simply by focusing on how they benefit the individual, if not directly then in the form of reproductive prowess. Evolutionary success is measured by the production of healthy offspring with genes similar to those of the parent. However, both before neo-Darwinism (in the writings of Darwin and other evolutionists) and after it (including here in *Chimeras and Consciousness*) the mathematically tenable but conceptually stunted notion that evolution is nearly exclusively a matter of individual survival is put to the test and abandoned.

The world, as we know and sometimes love it, is not open to conquest by one individual or one species. The history of life is punctuated by environmental crises, dwindling reserves, and even threats from itself that required concerted group action. As materially and energetically open complex systems, we naturally join forces, goals, and genes. We merge chromosomes and chemistries within species, bodies, and minds.

The word "chimera" refers to a mythical beast that blends the parts of real animals. The word thus has two connotations:

1. something questionable, mythical, imaginary ("chimerical")

2. something that represents the successful merger of two or more distinct beings.

It also has some specific meanings in various fields; for example, in animal genetics it refers to an individual with genetically distinct cells (e.g., one that results from the merger of one egg with two sperm that display distinctive characteristics, such as differently colored skin in neighboring tissues, in an adult).

In early cell evolution, two kinds of bacteria merged (technically, "eubacteria" with archaebacteria = "archaea"; see appendix A). The wholesale fusion formed more complex amoeba-like swimming cells (protoctists). One member of the two prokaryotic domains that merged probably was a sulfide-gas-making archaebacterium. The other, the eubacterium, most likely was a swimmer that oxidized sulfide gas and deposited its product: elemental sulfur. Together, as a permanent merger, this ancient chimera in the mid-Proterozoic Eon (c. 1,200 million years ago) became the earliest nucleated cell (figure I.1, right). The swimming, sulfur-metabolizing chimera was an ancestor to all nucleated life forms alive today. Not only did all animals and plants descend from the swollen, swimming chimera; so did the other protoctists and the fungi (see appendix A).

As you receive this message, courtesy of a plant pulp intermediary (i.e., a page of a book) or, less likely, glowing minerals assembled from elements in Earth's crust (i.e., a computer screen), the other organisms and you interact, ally, augment, differentiate, pool skills, make sacrifices, and sometimes, as in the case of the formation of superorganisms (global human society included), begin to merge again.

Chimeras are real. Life is not shy. Evolution can be violent and competitive. But cells and organisms behave in the context of their histories, their societies, communities, and bodies. They share. So too does our book. Though incomplete, it overflows with connections and potential connections as it chronicles the farthest reaches of group living.

In the first chapter in part I, Frank Ryan, a doctor of medicine, a symbiosis specialist, and a best-selling author, introduces us to viruses—biological entities so small they are not even selves. Ryan shows how the viruses we have been taught lead to disease, immune-system collapse, and maybe even death have undeserved bad reputations. All viruses need to be inside live, active cells before they can do anything. Outside cells they are like salt crystals: alone they do nothing at all. But many viruses, once inside a live cell, actively help ensure the continuity of health. Some viruses enable bacteria to provide us animals with nutrients, and many viruses have deposited their genes in our bodies. Many serve to cull dangerous over-population in ways that eliminate the weak or aged and promote and invigorate young and strong reproductives. Like most of this book's short chapters, Ryan's is dense, scientific, and provocative.

In chapter 2, William Day, a biochemist and an expert on the origins of life, traces bacteria back to non-living growth processes of chemicals energized mainly by sunlight. His detailed analysis breaks new ground by combining well-established chemical specifics with a metabolic over-view. According to Day, the entire growth phenomenon appeared simultaneously, with the earliest life. Growth, present from the onset, never ceased. Life did not originate with the "primordial soup." No prebiotic cake mix jumped around and reproduced as soon as "prebiotic DNA" formed outside in the environment. No DNA entered the lucky mix and "sparked" it. In chapter 3, Antonio Lazcano, Arturo Becerra, and Luis Delaye broach life's earliest sensation with their story of the importance of alarmones, analogous to plant and animal hormones. These bacterial chemical signals produced by distress help shape microbial communities. In chapter 4, Ken Nealson, a microbiologist long involved in NASA's search for life beyond Earth, shows how sensitive bacteria are. They detect mechanical stimulation, light, and chemical and energy gradients. They swim toward and away from high light intensities, low oxygen concentrations, or toxins. They alter one another and form groups that differ from the individuals that constitute them. Vast populations of mixed kinds of bacteria dramatically and persistently alter their immediate environment.

In part II, groups—life's components—join in metabolic and information-processing unions. From groups, emergent capabilities evolve. Both communication and social intelligence are found in bacteria that chemically interact. Eshel Ben-Jacob, Yoash Shapira, and Alfred Tauber outline how social bacteria—ones with all the same genes from a single-common ancestor—sense dwindling resources and respond. The bacteria survive hard times. They form quiescent stages, many even boil-proof, called spores. Some even remember that they suffered environmental insults, close encounters with antibiotics.

We pride ourselves on finding great real estate on lake shores, near beaches, and other coastlines; we prefer to live at slightly elevated altitudes, at places near clean water that meets land and air. We build spectacular homes insulated from strong winds and drenching rains. But such practical and beautiful architectural design preceded people by at least 2,000 million years. In chapter 6, Wolfgang Krumbein and Celeste Asikainen show us landscapes still formed in part and occupied by microbial masters. At the sea or in shallow lakes in houses of their own making, some bacteria and certain algae construct their own dwellings. In a specific repeatable sequence, community members build structures inside which they live. Some work metals, such as iron and manganese. Bacteria and other predecessors paved over and claimed choice property long before any vertebrate walked or swam the Earth.

At a personal level we can relate to biological groups in general, but one chapter on psychology and evolution will especially surprise us. In chapter 7, Laurie Lassiter introduces us to "Bowen theory." She shows us how human beings habitually behave in the context of their early family life. Autonomy is always compromised, along with free choice, as the reproductive needs of one's own group—whether a family, a tribe, or a chain of differentiating cyanobacterial cells–are accommodated.

In chapter 8, the evolution geographer James MacAllister reviews the state of interdisciplinary appreciation of group actions in natural communities through geological time. He concludes that the role in evolution of groups—chimeras, ranging from symbiotic partners to the largest ecosystem, Gaia—has been dangerously underemphasized. Gaia is the name for the physiological system at the surface of our planet. Earth's life-steeped surface, taken as a coherent living entity embedded in its watery environment and recognized as a single system, is no Goddess, it is Gaia. In this single system of composite ecosystems, temperature, atmospheric gases, and other changing, cycling environmental variables are regulated much as in an individual mammal's body. (This is not the same as saying that "Earth is an organism"—a poetic, New Age, but unscientific version of Gaia.) No organism entirely recycles its gaseous, liquid, and solid waste by itself. Rather, Gaia resembles a superorganism. Better yet, Gaia is much more an enormous set of nested communities that together form a single global ecosystem than she is any single organism.

In the first chapter of part III, the rational astrologist and student of biological clocks Bruce Scofield discusses the rich and varied influences of our cosmic cyclical environment, including the Sun, the Moon, and the planets, on our biological rhythms. In chapter 10, the long-time NASA geologist Paul Lowman and the classical composer Nathan Currier argue that plate tectonics, which strikingly differentiate our living orb from the dead spheres of Mars and Venus, is made possible by liquid water. Water, cycled and retained here on our planet by life, shaped this planet's surface. In chapter 11, Peter Warshall, an ecologist and a former editor of *Whole Earth Review*, traces the evolution of life forms that sense, respond to, and exploit various wavelengths of life-sustaining light.

Part IV opens with a short statement by Victor Fet, a Russian-born naturalist, poet, and expert on scorpions. He introduces us to his former countrymen's early contributions to the science of symbiogenesis. That mergers of different life forms are sources of evolutionary novelty is a well-known concept in Russian history of biology. Mereschkovsky's original insight is that photosynthesis (of all algae and plants) began as undigested food. The acquisition of photosynthetic organelles is owed to permanent indigestion. Such well-expounded concepts of symbiosis were ignored or actively denigrated, especially in Western countries (Khakhina 1992; Kozo-Polyansky 1924).

In chapter 13, John Hall and Lynn Margulis present exciting new evidence for and an interpretation of life's deepest, most potentially profound level of chimerical union. This ultimate story of symbiogenesis is under investigation now. Squirming sulfide-seeking spirochetes enjoyed unprecedented successes as they united and fused with anaerobic archaebacterial partners in sulfur-rich muds of yesteryear. Such mergers formed macroscopic life's most enduring—and evolutionarily pregnant—union. The idea is directly and easily testable in animals. The spirochete evolved into the cilium (Wier et al. 2010). The sensitive nucleated cells, arranged into specialized organs, such as eyes, taste buds and inner ears, today form the basis of animal sensitivities, if these authors are correct.

It has been said that if all the DNA in a single human being were stretched out end to end it would reach to the Moon—more than 230,000 miles. In chapter 14, the research biologist Andrew Maniotis, whose Chicago laboratory takes apart and puts back together normal and cancer cell genomes to figure out how the latter divert the body's resources to their own nefarious purposes, explains how the bacterial genophore —a microscopic DNA circle or in some cases a straight line—evolved into animal (including human) chromosomes. How do miscues in the timing of growth and duplication of chimeric chromosomes correlate with the punctuated evolution of the ancestors of the peculiar primates, the lemurs, that, after 60 million years, dwell only on the isolated island-continent of Madagascar? In chapter 15, the zoogeographer and mammalogist Robin Kolnicki tracks the most probable way.

Our attention is drawn to yet a further level of the importance of chimeras in evolution in chapter 16, where the biology teacher Sonya Vickers and the marine biology research scientist Donald Williamson concisely explain Williamson's theory that some of Earth's most beautiful and intriguing life forms are chimerical at the level of entire animal genomes. Fertile sex, it appears, has occurred between members of different animal phyla. Such interspecies mating in animals other than mammals appear to occasionally have been spectacularly successful. By hybridization they produced larvae, such as the caterpillars that are entirely different from adult butterflies.

In chapter 17, the physiologist Margaret McFall-Ngai advances a surprising chimerical hypothesis for the origin of the immune system. Bacteria, over millions of years, did far more than cause animals to be ill or to die. In chapter 18, the paleontologist Jessica Hope Whiteside and the writer Dorion Sagan delve into the symbiotic and evolutionary bases of human health. The medical subject is recognized not as an individual biological island but rather as a co-evolved community of interacting cells and organisms of distinct origin.

Part V, titled "Consciousness," begins with a comparison of brains. In chapter 19, the behavioral physiologist Gerhard Roth questions the attribution of "higher" human cognition just to the brain's hypertrophied cortex, since this part of the vertebrate brain also is enlarged in simians and in cetaceans. Roth emphasizes our high number of cortical neurons and the speed with which they process data. Other life forms, as we see throughout this book, are self-aware, but complex syntactical language appears to be unique to us humans.

In chapter 20, the neurophilosopher John Skoyles develops the concept that the human brain has evolved to the point that changing symbols are integrated into brain neurobiology. This development led to notable evolutionary consequences. In chapter 21, Dorion Sagan traces our purposeful choice-making behavior to the directional bias of naturally occurring complex systems ruled by thermodynamics. In chapter 22, the primatologist Judith Masters links chimeras and consciousness in yet another way by showing how acoustic and olfactory signals are used in primate troops and families for tracking and communication. In chapter 23, Luis Rico, a cultural historian and an artist, emphasizes the need to transcend disciplinary barriers. We live together on an indivisible, living planet within a cosmos of information processes that we have only begun to fathom. Evolution of life for thousands of millions of years is a story far more nuanced and exciting than simple accumulation of random mutations or a bloody struggle for existence.

Come with us, then, to survey, with exciting details, the long history of community and the recent shock to the planet caused by the emergence of the mammalian body-mind. We supposedly wise *Homo* are 6,800 million strong and seem to believe that Earth can accommodate our unceasing population increase. It cannot.

The reproductive potential, scarcity, hunger, thirst, contention, suspicion, love, merger, fecundity, and mortality that produce evolutionary novelty produced us. We emphasize awareness. Like other animals, and indeed all other life forms, we humans do not live alone. We have always been embedded in communities that include our viral, bacterial, protoctistological, fungal, and plant traveling companions. We continue to ignore them at our own peril.

This book is designed to put us upstart primates in the sensory context of other life forms. Since early bacteria avoided oxygen, and the first spirochetes wriggled through organic ooze toward sugars, sensory systems have expanded and complexified. Even cognition, the reorganization of sensory input toward the emergence of meaning, may be bacterial. Bacteria were first: their unprecedented activities in cell reproduction, genetic recombination in sex, networking, photo- and chemoautotrophy, methano- and sulfidogenesis, seaside architecture, air pollution, and other activities changed Earth's surface long before the evolution of the first marine animal. Communication among millions of life forms (strains, varieties, and species) has been rampant nearly since the origins of life itself. Bacteria detect Earth's magnetic field and gravity. Protoctist bodies evolved eyes, tentacles, and fishing rods to help catch prey. The biosphere is abuzz with more-than-human sensation and information flow. Chemical communication among trees, whale sonar systems, and, more recently, people who talk, read, and write electronically have augmented the non-stop tendency of this life to reach out to other life. Communication modes that began in crowded bacterial mats and scum have been in place, grown, declined, and changed for at least 3,000 million years.

Membranes, ion channels, and electron flow in and out of wet cells provide a basis for communication in all forms of life studied. Ching Kung, a professor at the University of Wisconsin at Madison, has pioneered research into mechanosensitivity: electric organs, temperature sensing, sound, and all forms of touch. The material and energetic basis of these senses involve water and salt (Kung 2005).

Kung has revealed molecular activities that correlate with predictable behaviors. His search to analyze simplified patterns of behavior began with the pond water ciliate *Paramecium*. He found that a single mutant, a single change in a base pair in DNA that led to a substitution of a different amino acid in a protein, corresponded to a single impressive inherited change in the total behavior of this large swimming ciliate.

The mutant organism is called a "pawn" because, like the chess piece, it only swims forward. Its normal parent swims forward, swims in circles, turns around, changes direction, swims backward, and in general enjoys a large repertoire of behavior relative to the pawn mutant. Kung's elegant experiments, and many others, cleverly established that the sensitivity of life to mechanical stimulation, to touch, is mediated by charged ions (for example, K^+, Na^+, or Ca^{2+}) in solution in the flow of water through protein channels in cell membranes. Since all life is cellular and all cells are self-bounded by membranes in which protein-ion channels are embedded, all life senses water, salt, and touch. Kung's amazing model of mechanosensitivity includes sensitivity to gravity and to magnetic fields (gravisensors and magnetosomes). Taken together, this implies that the detection of movement of water and salt preceded even the first prokaryotic cell—that is, our last common ancestor (LCA), also called the last universal common ancestor (LUCA). Some sort of early bacterial cell sensed sunlight, other starlight, and moonlight. Its descendants still do.

That tiny ancestor attracted nutrients (chemosensitivity as semiochemicals, taste and smell) and was repelled by noxious acids, heavy metal ions, and other salts. Recognition of the subtle sensitivities of our bacterial ancestors leads us to consider how pressure changes, sound, water and other waves, heat, light, and other stimuli work through our sensory systems to organize expanding selves and their behavior.

All life, as far as we know, resides within about 20 kilometers of our planet's surface. From 8 kilometers to the top of the troposphere to about 12 kilometers to the bottom of the ocean at its deepest point, this is the "Biosphere." Vladimir Ivanovich Vernadsky brought the extent of the biosphere and its importance as the abode of life to the attention of modern scientists for the first time in 1926. Although a French translation of his book (titled *La Biosphère*) was available by 1929, Vernadsky was unknown to the English-speaking world until the end of the twentieth century—and still is (Lapo 1987; Vernadsky 1998).

Living beings reside in habitats, portions of ecosystems that they incessantly sense. The sensory physiology connecting beings to their immediate surroundings is ubiquitous. Even dormant seeds, drifting spores lofting on air currents, and mammals sleeping in subterranean burrows sense and respond to their habitats. Habitats are enveloped by specific ecosystems. Any ecosystem is surrounded by the rest of Earth's great biome (Folch et al. 2000). Any particular organism is always a part, a minuscule part, of the living environmental regulatory system at Earth's surface—that is, of Gaia (Lovelock 1988).

Earth's atmosphere and surface sediments are regulated, at least in part, by the activities of the more than 30 million types of living organisms that live here (Crist and Rinker 2009). Earth is soaking wet relative to its planetary neighbors and enjoys a radically energetic, reactive atmosphere relative to the boring carbon dioxide worlds of Mars and Venus. Earth, so far as we know, is the only abode of life in this universe. So perhaps the most inclusive answer to the question "What does life sense?" is "Itself."

Gaia is thirsty. The average depth of water is 3 kilometers (3,000 meters). On Mars, the average depth of water (vapor and liquid) is about 0.001 meter (1 millimeter). On Venus it may be a few centimeters. Whatever the precise amounts, we conclude that Venus and Mars are dry deserts relative to Earth. With respect to its neighbor planets, Mars and Venus, Earth is a "water anomaly" in the solar system. Scientists and other people tend to assume that environments are "physico-chemical givens" to which "life must adapt" in order to survive. Not so. Going against the prevalent belief that "life adapts to its environment," or, put another way, passive life is a mere passenger on Spaceship Earth, Gaia theory posits that life actively regulates biologically relevant aspects of Earth's surface. Life forms strive to maintain habitable local conditions. Gaia theory specifically proposes that water regulation emerges from life itself (Harding and Margulis 2009). Earth's organisms incessantly respond to their abiotic and biotic surroundings: temperature change, chemical composition of air, soil, and water, light and dark conditions, mechanical impediments (wave actions, wind, rain, snow, fire), and many other variables. And response to water, both thirst and satiation, seems universal.

Life's sensitivity to the quantity and the saltiness of water may be the most elemental of all senses. Plants, animals, fungi, and all microorganisms know thirst and assess wetness. Life senses local water levels. The universality of water detection and the response of living cells to this ubiquitous solvent seem to reside in the properties of the lipid-protein membrane. This membrane, called a "semi-permeable bilayer," is the intact external boundary of all cells at all times. When it is breached and its integrity is lost, the cell, whether it is a small bacterium or a large egg, dies. Once lost, the ability to regulate the flow of material and energy through the lipid bilayer of the cell's membrane is irretrievable. Water leaks out. The permanent one-way loss of water out of the ruptured membrane is a symptom that the food (material) and energy transformation have stopped. The thermodynamic system of life halts (Schneider and Sagan 2005). This has a name: we call it death. As soon as the self-maintaining activity of an organism ceases, its material will be replaced by an inert puddle of carbon-hydrogen-nitrogen compounds. Identity of a life form, whether of one or more cells, requires a continuous influx of energy and material. In the absence of self-maintenance, any being immediately loses all signs of animation. Formerly alive matter transforms. It becomes food for those that retained their intact membranes and, with them, the profound sense of water.

Water sensing requires at all times the ubiquitous lipid-protein cell membrane that confers identity. "Ion channels" that sense and move salts are embedded in it. Most surprising to us is the set of facts established by Ching Kung and his colleagues: fragments of membrane taken from living cells retain properties of sensing and response to salt and water for quite a while after the cell has died. For minutes, even hours, Kung and his colleagues "poke" isolated membranes and measure behavioral activities. Specialized proteins embedded in the outer membranes of cells recognize familiar "minerals"—that is, ions in solution, such as calcium (Ca^{2+}), potassium (K^+), hydrogen ions (H^+, acidity), and hydroxyl ions (OH^-, alkalinity). The membrane-channel-making proteins can block and release the flow through them of ions in water. Some of these channels are activated by a familiar force: touch.

Mechanical stimulation, the simple bending of the membrane lipid bilayer, blocks and/or releases the flow through it of ions in water. This block and release, underlaid by the closing and opening of the channel by protein shape changes, responds to various stimuli. Certain ion channels are sensitive to mechanical stimuli, including touch, blood pressure, sound, and changes in salt concentration. A *Paramecium* recoils from physical impact at its head end. Research also shows that pressure exerted on a patch of membrane can stretch it. This stretching causes certain bacterial channels to open or close. It thins the lipid bilayer and creates a chemical change that causes the channel to open. For living bacteria, an increase in osmotic pressure when water rushes into the cytoplasm after a rain causes the same membrane stretching. The fact that organisms sense water and show many behaviors that help retain it in their immediate vicinity has had a profound effect on the collective of life. We think that the continued presence of water on Earth for more than 3,000 million years must be attributable to life's activities. Life's presence has depended on the continuation of planetary water, not on a lucky place in the sun.

Life does indeed change its environment to fit itself, as L. J. Henderson and Ian McHarg insisted. (See McHarg 2006.) But does the assertion that "any organism" is "well adapted to its environment" have any meaning? All organisms alive today are "adapted" by virtue of the fact that they live. Their ancestors have survived from the past to the present. Gaia emerges as different organisms affect one another and their sur-roundings. Through the exchange of heat, light, liquids, gases, and a huge array of chemical elements, their salts, sugars, long polymers like starch or cellulose, and other chemical compounds, the "creative fitting of health" can be documented. McHarg notes that the trouble and pain of an unhealthy person, family, woodland, or neighborhood is recogniz-able as the "reductive misfit revealed in pathology," in which the organism-environment fit is threatened. The biological reality of health, in contrast, depends on incessant effective communication between the live organisms and their physical and social environments via senses.

This book proffers a modest, tentative beginning of a scientific inquiry into the evolution of the collective sensory capacity of life. In it we ask what it is that underlies our subjective sense of human consciousness. Is not this awareness of ourselves and others a reality we feel is true? Do we not agree that the traditional scientific emphasis on "objective reality" and on the need for "incontrovertible evidence" and "absolute truths" tends to miss what it is about "consciousness" that most interests us?

We hope that this book just begins to make amends.

I

Selves

1

Valuable Viruses

Frank P. Ryan

As the positively spun notions of "going viral" and "viral advertising"—spreading something useful or exciting in the media—suggest in the human realm, so we are learning that, in the deeper realm of biology, viruses themselves cause far more than sickness. In fact, they appear to be evolutionarily crucial, carrying genes among cells and helping to create new structures, processes, and organisms. Here Ryan discusses how viruses, although not autopoietic entities nor alive, promote healthy life by engaging in "cyclical symbioses."

Wherever they have been sought, either in prokaryotes or in eukaryotic life forms, viruses have been found. They are insensitive. Unlike bacterial or nucleated cells, viruses alone cannot reproduce, metabolize, or grow. Nonetheless, the role of viruses in evolution has been grossly underestimated. Their dismissal as "dangerous germs" is more a reflection of our ignorance than of our understanding of their deep involvement in group living arrangements. Viruses have one or another type, not both types, of long chain nucleic acid molecules (DNA or RNA). They enclose themselves in protein coats called capsids. They contain genes for some enzymes, such as integrase, RNA, or DNA polymerase, but have fewer than a cell's worth. The smaller viruses have from two to four genes; the largest may have hundreds. The "system's properties," the "selves" of viruses, utterly depend on their physical contact with bacterial or other living cells. If not connected to a cell, a virus is as inert as a lump of salt or a cube of sugar. The basic element of life, the self, is the sensitive bacterial cell; but a virus, as a courier and an integrator of genes into bacteria and nucleated organisms (animals, plants, fungi, and protoctists), can be very important to specific evolutionary trajectories (figure 1.1).

Virologists have become increasingly aware of the potential of viruses to influence the evolution of the cells and organisms in which they reside.

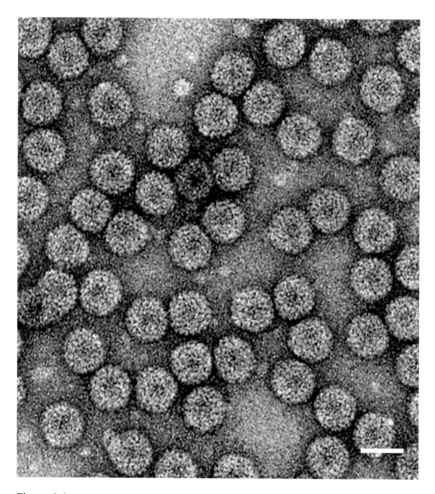

Figure 1.1
Reverse transcriptase virus. Scale bar = 0.1 μm.

All viral relationships depend on the active metabolism of the cells in which they dwell. Viruses alter the evolution of the organisms in which they reside—bacteria, protoctists, animals, plants, and fungi—by incorporating their genes into reproducing cells. The result is not, as many would be likely to think, always diseased organisms or cells. Sometimes chimeras—novel "holobionts" (Ryan 2007)—arise.

What are the general principles of virus-organism interactions? We must study "cyclical symbioses": how viruses move genes into and then out of living cells and whole organisms. Viruses have profound effects on populations of organisms; in the end, all organisms live as members of their populations. We need to understand the principles of viral-organism interaction that enable us to extrapolate the potential for future research.

An infectious virus can change mammalian genomes through a process I call "aggressive symbiosis with plague culling." In essence, a plague virus will cull the host population's gene pool of genotypic variants that are unable to live with the presence of the virus. The evolutionary implications focus on genetic pathways involved in the immediate animal viral interaction, notably those that affect viral lethality and control of the immune system. For example, the rate of progression of AIDS is strongly associated with a particular HLA-B but not HLA-A immune gene expression. This affects the evolution of the HIV-1 virus. Meanwhile, HLA-B gene frequencies in the population are those likely to be most influenced by HIV disease. Other purported effects of culling in mitochondrial and major histocompatibility genes of chimpanzees and hominids were described by De Groot et al. (2002) and by Gagneux et al. (1999). More profound and varied symbiotic contributions are associated with persistent rather than acute virus-mammal interactions, where infection may be via the parents to the offspring ("vertical" transmission) as well as by direct transmission from member to member in the same population ("horizontal" transmission). The interaction and evolutionary dynamics in persistent (chronic) viral infection differ greatly from those in acute transient viral infection. The latter is more familiar because it manifests as disease; the former is more important in terms of the viral influence on animal evolution. In chronic viral "infection," viruses enter permanent relations with mammals such that the animal does not eliminate the virus from its life history but the virus no longer damages it. These associations lead to the development of stable long-term relationships (Villarreal 2007). The result can be persistent, intimate, and sometimes beneficial interactions between the animal and its virus infection. Similarly, beneficial long-term interactive partnerships arise between viruses and bacteria cells or between viruses and protoctists.

What do persisting viruses bring to cyclical symbioses? Aggression is sometimes a behavioral contribution of viruses to the holobiont. Some 25,000 species of parasitic wasps enter partnerships with approximately 20,000 types of polydnavirus. The virus may be incorporated into the wasp genome, but may also be transmitted parentally (vertically) with no wasp-virus genomic integration. The interactions may be more complex. Aphids, which are prey insect species for some of the wasp-polydnavirus associations, enter into obligate partnerships with bacteria, notably buchneras. The Gram-negative bacterial symbionts supply the aphids with nutrients such as nitrogenous compounds and other nutrients missing from their diet. In addition to buchneras, some species of pea aphid harbor additional facultative bacterial symbionts, such as *Hamiltonella defensa*, that behave aggressively. These infect and kill the larvae of the symbiotic wasps. In turn, *H. defensa* depends on a virus, called bacteriophage APSE-1, that facilitates bacterial penetration of the insect's cells.

Closely related phage viruses are important to a wide range of insect-bacterial symbioses that require the bacteria to invade animal cells (Moran et al. 2005). Where viruses are incorporated into the nuclear genomes of germ cells, this generates a great potential for interaction at the genetic level. Viruses are remarkably creative in fashioning new genes by amalgamating genetic fragments. Thus, it comes as no surprise that most viral gene sequences are not found in either the prokaryotic or the eukaryotic databases. Viruses are recognized to have unique capabilities in manipulating mammalian and other genomes, as in viruses of the centromeric DNA of the chromosomes of interspecies rock wallabies. There the viruses cause breaks (fission) and reconnections (fusion) between chromatin sequences in chromosomes. This process, known as "chromosome juggling," massively accelerates adaptive radiation that generates new species "instantaneously" (from a geological point of view) (O'Neill et al. 1998).

Cyclical symbioses that require the transfer and incorporation of prokaryotic or eukaryotic genes between symbionts are likely to depend on viral genes. For example, the nodulation and fixation genes of the rhizobium *Mesorhizobium loti* are encoded in the bacterial genophore as a 500-kilobase "symbiosis island" that requires a P4-phage enzyme, known as integrase, for transfer of the symbiosis island from symbiotic to non-symbiotic strains of rhizobium (Sullivan and Ronson 1998).

The symbiosis that involves sequestration, continuing viability and photosynthetic activity of the yellow-green plastids from the golden yellow alga (the xanthophyte, *Vaucheria litorea*) to the sacoglossan mollusk, the sea slug, *Elysia chlorotica*, apparently depends on an as-yet-unclassified retrovirus. If so, the likely mechanism would involve the transfer of genes from the nucleus of the alga to the sea slug (Rumpho et al. 2008). This RNA-dependent virus, found in prodigious numbers in the tissues and the nucleus of the sea slug, prevails in degenerating adult slugs and on the surface of seasonally laid eggs.

Viruses are universal and transmit specific genetic information across species and more inclusive taxonomic boundaries. Their potential to integrate bionts in symbioses is extraordinary, relatively frequent, and often overlooked. Marilyn Roossinck's pioneering application of sym-biological methodology to virus-fungal interactions resulted in the dis-covery of hitherto unknown viral symbionts inside the fungus *Curvularia protuberate*, which is symbiotic in grasses that grow in especially hot dry conditions. Indeed, the prokaryotic genomic fusions implicit in Serial Endosymbiotic Theory (SET) suggest that viruses may have been involved in the specific mergers hypothesized in the origins of undulipodia (kary-omastigonts), mitochondria, and plastids. (See chapter 13.) The recent finding of specific centrosome-associated RNA sequences (cnRNAs) with at least one conserved reverse transcriptase domain is a tantalizing sug-gestion of a viral role in centrosomal-nucleolinus evolution (Chapman and Alliegro 2007; Alliegro et al. 2010). Here, as in other avenues of exploration, recent insight into viruses and their potential in symbiosis as part of "self" rather than "foreign agents" provides exciting new pos-sibilities for research.

Viruses are observed in cells of all major taxa, from archaebacteria to charismatic zoo animals. Since they lack metabolic pathways and self-maintained lipoprotein membranes, viruses, though not "selves," are still fundamental to genetic fusions across disparate taxa lineages and in the origin of sensation in cells and organisms, particularly in the integration of free-swimming bacteria as they become trapped as organelles. Viruses are still involved in the development, maintenance, and evolution of holobionts today.

2
More Like a Waterfall

William Day

As mysterious and complex as life is, essentially it is a sustained growth process that depends on energy flow and matter. Understanding this illuminates how it originated and evolved. Here Day proposes logical, detailed steps for the chemical unfolding of life growing in its energetic milieu.

Life, both in origin and in evolution, is more a process than a thing. We need to understand what life is—a natural biochemical "growth" process—to interpret and design experiments. Identifying life as a sustained growth process greatly simplifies deciphering how it originated and evolved. Although many origins-of-life experiments have produced complex organic compounds such as amino acids from gas mixtures taken to be representative of Earth's early, Archean atmosphere, such chemical products represent only a small piece of the puzzle. I argue that light induced the emergence of the cell's constituent systems and stabilized its dynamic states. The laboratory-generated carbon compounds produced by electrical discharge or other energy sources of traditional origins-of-life experiments are reminiscent of the biblical story of Genesis (Hazen 2005). More complex compounds result from sparking methane, ammonia, and water vapor. The original Miller-Urey experiments (Miller 1953) started the "origins of life" chemical studies. Harold Urey, a Nobel Prize recipient, wanted his graduate student Stanley Miller, then in his early twenties, to be given credit for the dramatic experimental results (Miller and Urey 1959). They and their successors produced formaldehyde, simple sugars, and hydrogen-cyanide-derived polymers, including the intracellular energy-storage compound adenosine triphosphate (ATP) in the absence of any life or even help from enzymes extracted from live cells. Other important biochemicals, including orotic acid, pyrimidine precursors, and lipids, were made by energizing gas mixtures. This was "origins of food," not "origins of life."

We have seen very little progress since these groundbreaking "prebiotic chemistry" experiments. The resolution of how life originated ought to be no more difficult than any other fundamental scientific inquiry. But to explore and explain life's origin properly we need to understand what life is.

First of all, we need to recognize the way in which living matter as a system differs from the usual substances treated by chemistry. Scientists believe life's properties come from the precise composition and organization of its components. They assume, therefore, that for life to originate, prebiotic substances must have assembled into structures of sufficient complexity for life to have arisen and begun to function. The assumption is that life is a special "thing" that appears, almost magically, once a certain threshold of complexity is reached. The first cells needed just the right mixture of ingredients, the exact right molecular mixture.

I disagree with this extremely improbable "cake mix" concept. I do not think that if proper proportions of life's ingredients (such as amino acids, lipids, and nucleotides) are brought together and mixed with an adequate energy source life will ever emerge.

The probability issues that plague this building block/assembly hypothesis disappear when we recognize that metabolism, reproduction, and evolution are aspects of the same growth process that developed over time (Margulis 2007). No pre-existing store of information was required for life's formation. As the emerging system grew sequentially, information accumulated in the structure and order of the reactants of the metabolic system. Only after the cell appeared was information coded in DNA, the nucleic acid of the cell's genetic system.

No single cell or living body is ever static; living beings dynamically produce themselves by regeneration. The living system continually breaks down its components by lytic enzymes and resynthesizes its parts anew, as Schoenheimer (1942) and his co-workers showed. Such chemical turnover is not equilibrium chemistry, but rather a steady state of metabolic reactions that continually consume and expend energy. When the energy and material flow that fuels buildup and breakdown is disrupted, for whatever reason, synthesis ceases. Matter and energy flow are absolutely required at all times. The unbalanced degradation reactions continue until thermodynamically unstable structures collapse and life is irreversibly lost. At absolute zero or in dormant propagules this imperative is suspended. As soon as energy and matter flow, the requirements for life are restored, and metabolism resumes.

Recognition of life as a process immensely simplifies and clarifies how it originated. In the flow of material and energy that defines life, process is inseparable from structure and composition. Biochemical synthesis, autocatalysis, and reproduction are aspects of a single growth process. Life is less like a clock assembled by an outside maker than a waterfall forming itself as it proceeds.

The simplest, most direct account of life's origin and evolution should follow the order in which life's systems could have developed. The apparent order in which these constituent systems evolved was as follows: energy-growth core, membrane, autocatalytic cycles, metabolic system, genetic system. When we follow this order of evolution, we can see that the cell developed in coherent continuous causality, each new system integrating and holistically stabilizing the previous one in the gradually emerging cell.

Popular scientific theories, in my humble opinion, neglect the need for a coherent and proper causal order of the minimal biochemical reactions that sustain life. Without a causal chemical sequence one may have food but no theory of the origin of life itself. This includes Gerald Joyce's RNA concept, where the fragment of RNA replicates, but then what? Without sustained energy and matter flow (i.e., life), the RNA is only fancy food. David Deamer's liposome ideas that life started as microscopic bags of fatty membrane-like bubbles that concentrate and surround other organic compounds and Günter Wächtershäuser's idea that life began on mineral surfaces have individual merits. Wächtershäuser's concept that life began through electron flow on mineral surfaces, probably sulfidic or iron (or both), is excellent. I agree, but he stops short of coupling the energy flow to the continuity of membrane-bounded continuous biochemical synthesis. Therefore his, too, is a variation of the "origin of food," not the "origin of life" process. Just because organic compounds can be synthesized in the laboratory and in nature without the presence of life does not necessarily mean these syntheses are relevant to the origin of life.

A Sustained Growth Process

Visualizing life as a sustained growth process, one is better prepared to infer the order in which the biochemical processes at the heart of its evolution proceeded. Central to life, both now and in the past, is the production of high-energy compounds found in all life today: organophosphates and sulfur-organic chemical linkages called thioesters. They activate reactants and make chemical conversions spontaneous. Repeated activation probably started with relevant ions (Ca^{2+}, Mn^{2+}) and divalent metal ion catalysis as a reaction series grew and produced complex molecules. Among the first organic compounds to be synthesized were likely those with catalytic properties that are now coenzymes. Their interlinking in mutual catalytic dependence created the metabolic system.

The core system, the universal metabolism of all life, was encapsulated in protocells—liposome-like bags of abiotic origin—or enclosed in fatty membranes, since the beginning of the system. The formation of the DNA genetic system with its complex protein synthesis activities probably occurred only after the membrane-integrated growth system was in place. A thought experiment shows how this is the case: if life were essentially nucleotide replication, independently evolved, why would relatively slow growth via metabolism integrated into cell membranes have evolved (Wicken 1987)? The integrated genetic-protein system evolved not in any "prebiotic soup" but rather inside the earliest minimal cell, what I call "the energy-growth core."

The Chemistry of the Core of Life

Three processes necessary for cell origins are activation of organophosphate reactants, synthesis of the bounding membrane that establishes identity, and incessant transfer of electrons (oxidation/reduction energy flow). To occur spontaneously, reactants must have a higher free energy than their products. If not already spontaneous, the reactant must be activated to a high-energy derivative. The cell needs a simple procedure to convert "low-energy" chemicals into "high-energy" activators. This procedure began, I suspect, with the dehydration of 2-phosphoglycerate. In this dehydration reaction a relatively unreactive low-energy phosphate ester is transformed into phosphoenolpyruvate (PEP), the high-energy reactive anhydride of phosphoric acid and an unstable enol of pyruvic acid.

This reaction is central to the energy-growth core of chemical transformations that led to subsequent coupled reactions in close proximity. Other reactive phosphates were generated. A second molecule of 2-phosphoglycerate converted to 2,3-diphosphoglycerate by reaction with phosphoenolpyruvate, a compound that is a cofactor in the isomeric conversion of 3-phosphoglycerate. Thus, 2-phosphoglycerate still forms a catalyst (2,3-diphosphoglycerate) for formation of itself. I assume that the 3-phosphoglycerate was and still is produced simultaneously with the 2-phosphoglycerate. The conversion increases availability of the isomer.

Other molecules of 3-phosphoglycerate probably accepted phosphate from 2-phosphoglycerate and produced highly reactive 1,3-diphosphoglycerate. Thus, two important donors of the phosphate group, phosphoenolpyruvate and 1,3-diphosphoglycerate, are generated by the energy-growth core to activate compounds for spontaneous reactions.

The Membrane

When phosphoenolpyruvate transfers its phosphate in an activation reaction, it simultaneously yields a molecule of pyruvic acid, an a-keto acid particularly vulnerable to decarboxylation by ion (e.g., Fe^{2+}, M^{2+}, and HS^-) catalysis. This loss of a CO_2 molecule of pyruvic acid with hydrogen sulfide present yields acetyl-SH, a thioester with a high-energy bond equivalent to the phosphate bonds of adenosine triphosphate (ATP). Thioesters today participate in ester synthesis, including those in complex lipids. I hypothesize that the earliest membranes in emergence of the cell, made of isoprenoids, were synthesized by thioesters.

Electron Transfer

Sulfhydryl ions, strong nucleophiles under H_2S-rich conditions, tend to substitute mercaptans for organic phosphate. Phosphoserine and phosphoseryl compounds convert to cysteine. The tendency of mercapto groups to compete with hydrogen sulfide for bonding to ferrous ions gave rise to iron-sulfur (Fe-S) clusters bound to organics. The cysteinyl residues modulated the redox properties of the Fe-S clusters that led to the ubiquitous redox cycles of cells.

Iron-sulfur clusters function at the active site in ferredoxins and in enzymes tend to be embedded in the membranes of respiratory and photosynthetic chains. Oxygenases, hydrogenases, nitrogenases, some non-redox enzymes such as aconitase, and many other proteins. These iron-sulfur (Fe-S) clusters in ferredoxin suggests that they were essential in the transition from the inanimate chemistry of the lower Archean eon to the earliest living selves (Rees and Howard 2003).

Ferredoxins retain iron-sulfur complexed with cysteine molecules joined by polypeptides in a high proportion of small thermodynamically stable amino acids. These polypeptides, via ferredoxin, augment the production of ATP by radiation, the reduction of CO_2 to pyruvate, and also nitrogen fixation. More ancient than nicotinamide adenine dinucleotide (NAD), ferredoxins are the ubiquitous reducing agents in extant cells. I posit that an early form of ferredoxin was involved in life's origin.

A Light-Induced Origin?

Pyrite (FeS_2) is a common iron-sulfur mineral with two disparate properties that, I hypothesize, together sparked into existence the first "energy-growth core." Pyrite led to the continuity lacking in Wächtershäuser's proposal. Pyrite itself is comparable in catalytic activity to coenzymes that bear iron-sulfur clusters at their active sites. Pyrite alone transduces sunlight to an electron flow.

I suggest that organophosphates present from the beginning included membrane phospholipids, phosphoglycerates (for the energy-growth core), and high-energy activators for spontaneous transformation. Phosphoric acid in volcanic waters adsorbed on pyrite surfaces converted pyrite to hydrogen sulfide that supplied electrons for the photovoltaic reduction, which simultaneously became the reactant in organophosphate compound synthesis.

Archean volcanic gas emissions included CO_2, N_2, H_2S, PH_3 (phosphene), and, into the water, phosphoric acid. Carbon dioxide and nitrogen were reduced as hydrogen sulfide was oxidized. This oxidation/reduction reaction was "waiting to happen." The electron donors and acceptors paired in oxidation/reduction photovoltaic reactions where pyrite transduced sunlight and energized a flow of electrons between them. As hydrogen sulfide oxidized, carbon dioxide was reduced, such that phosphate on the surface of the pyrite reacted.

Synthesis of 2-phosphoglycerate in the absence of enzymes would verify these ideas. The condensation of formaldehyde from reduction of CO_2 and phosphorylation (esterification) in the presence of Mg^{2+} under conditions of photovoltaic reduction and surface adsorption ought also to be tested. Although these reactions occur separately under controlled conditions, I propose that the initial synthesis of the entire "energy-growth core" was the single process that has been in continuous existence.

Early Amino Acids

Chemical energy was carried as phosphoric acid anhydride as the life-process was sustained and continued to grow by incorporation of nitrogen into its compounds. Synthesis of a-keto-acids and their amination by NH_4^+ produced protein amino acids. Four amino acids, all synthesized directly from core reactions, are precursors to all other protein amino acids—serine, glycine, alanine, and aspartic acid. These are also the amino acids from which purines and pyrimidines are synthesized.

The first coenzymes of the energy-growth core of phosphoenolpyruvate, thioesters and ferredoxin (Fd) were ATP, acetyl coenzyme A, and NAD. They were incorporated into the energy-growth core system in that order. Synthesis of purine preceded that of ATP. It is no coincidence that the first amino acids (glycine, aspartic acid, and glutamine), along with formate and carbon dioxide, form the components of the purine skeleton. From the beginning the energy-growth core was correlated with processes that, inside membrane-bounded metabolic systems with continuity, became RNA, and then DNA polymerization reactions. The energy-growth core system that became the earliest cells appeared before nucleic acid (RNA, DNA) polymerization reactions.

Early Coenzymes: ATP and CoA

Adenosine triphosphate (ATP), coenzyme A (CoA), and nicotinamide adenine dinucleotide (NAD) are purine coenzymes. Since the biosynthesis of purine nucleotides today begins with ribose 5-phosphate, and since in an ATP-dependent pyrophosphorylation 5-phosphoribosyl-1-phosphate (PRPP) is produced, I deduce that this compound (PRPP) is the precursor to many other products, including the purines and pyridoxal phosphate.

Coenzymes are either nucleotides or analogs of nucleotides. This astonishing fact is a revelation concerning the origin of the minimal energy-growth core system: the bacterial cell. It shows that coenzymes used for the original energy-growth core could later provide replication components for stable patterns of nucleic acid synthesis. Four of these coenzymes—NAD, NADP, coenzyme A, and flavin adenine dinucleotide (FAD)—also contain an adenylic acid residue. Many have a nitrogenous base at one end of the molecule and a phosphate group on the other, with or without a small carbohydrate hinged between them. Alternatively, these two nitrogenous bases and phosphate joined in the form of a dinucleotide. The phosphate carriers NAD and NADP (the nucleotide adenine diphosphate) are true dinucleotides but contain nicotinamide instead of purine or pyrimidine.

One-half of coenzyme A is a nucleotide, although this is not its functional group. Thiamine pyrophosphate is a pyrimidine nucleotide phosphate in which a thiazolium ring replaces a sugar. The phosphate of pyridoxal attaches directly to the base. Once synthesized, CoA became the thioester of the lipid membranes. Acyl CoA activated by reaction with carbon dioxide formed the malonyl derivative. Malonyl CoA, the active form of acetate, generated fatty acid synthesis essential to the formation of membranes.

Coenzyme A contributed to the expansion of syntheses from the energy-growth core reactions to more complex biochemical pathways. It made possible chlorophyll-porphyrin and isoprenoid synthesis in the formation of the membranes of photosynthesis. With the formation of the intermediary metabolism that begins with the reduction of carbon dioxide in autotrophy for amino acid synthesis as proposed by Harold Morowitz et al. (2000), the citric acid cycle was incorporated as a part of the energy-growth core mode of synthesis. The citric acid cycle was used in prokaryotic synthesis long before it generated ATP energy in any mitochondriate eukaryote (protist or animal).

The first prokaryote with only a few amino acids in its energy cycle of high-energy phosphate compounds hypothetically established the first cell metabolism. The synthesis of coenzymes led cells to greater complexity. With only a dozen distinct compound classes, the coenzymes carried energy, catalyzed reactions, and transported electrons. A result of system continuity, the coenzymes with amino acids and other ambient precursors became a complex array of interrelated components. From the autocatalytic reproducing of the energy-growth core, the metabolic cyclical system originated as cells, which have persisted through time.

The Earliest Selves

The preceding section detailed a chemically logical progression leading to the earliest cells, the earliest life forms that involved energy metabolism and matter flow. Genes and their proteins evolved inside such early metabolic systems: bacteria or prokaryotic cells. The first metabolic entity on the lineage to the last (universal) common ancestor (LUCA = LCA) evolved into the first bacterial cells. The synthesis of life's characteristic macromolecules, proteins and nucleic acid polymers, was probably the last major biochemical core innovation to evolve. The modern genetic system, with its DNA, messenger RNA, and protein synthesis, evolved by the same step-by-step energy-driven material-flow metabolic growth process. ATP, nucleotides, and protein amino acids, none of which are produced directly from genes, were required.

By the time the minimal energy-growth core biochemical cyclical process assured itself continuity by the addition of the universal macromolecular modern genetic system's biochemical process, life, as the prokaryotic cell, had originated.

The cell's interior was a rich milieu of interacting chemicals, protected from entropic dispersion. It is likely that no new amino acids were incorporated after proteins were synthesized (Westheimer 1987). Life's origin would have occurred after all twenty of the protein amino acid pathways had evolved.

The four nucleotides that constitute DNA—adenine, thymine, guanine, and cytosine—carry energy when phosphorylated (adenine in the form of ATP, for example). They are the monomers that synthesize carbohydrates (UDP), lipids (CDP), and proteins (GDP). The metabolic core, beginning with the reduction of CO_2 and N_2, probably evolved in response to changing environmental gas supply as H-rich gases (NH_3, H_2, CH_4) were replaced by more neutral ones (CO, CO_2, N_2). By contrast, the genetic system probably emerged, as in Ernst Mayr's idea of "preadaptation," from nucleotides and amino acids that were already inside cells. Nucleotides were, as they still are, continuously synthesized. Oligomer condensation still occurs readily under the concentrated conditions in membrane-bounded cells. Dimers and trimers prevailed, although longer polymers appeared from their occasional coupling. Oligomers of six or more monomers sufficed to join in the evolution of the growth of double-stranded RNA.

The genetic system originated in this highly favorable environment within the enclosure where metabolism provided a supply of activated D nucleotides and the full complement of all twenty amino acids. As the synthesis of enzymes evolved, the genetic system and metabolic systems became mutually interdependent. The use of nucleotides and amino acids by the nascent cell provided it with enzymes. Metabolism preceded the genetic system. From the beginning the growth system depended on concentrated resources of both energy and matter. With time, metabolic and genetic systems integrated fully and the modern bacterial cell—LCA—was born.

Conclusion

Let's review how the current detailed view differs from vaguer, if more popular, scenarios for life's origins. Far from requiring an absolute level of complexity to begin, the growth process leading to life was initiated and sustained by light and electron transfer of organic chemicals run by energy flow. As the membrane-bounded cell with gene-mediated metabolism evolved, it sustained the inexorable exigency of growth. Reproduction emerged as a response to autopoietic maintenance and repair obligatory to the continuity of survival.

Chemical compounds synthesized in place are activated, reacted, and activated again; they alternate between being products and reactants. Electron and chemical transport occurs not by movement of molecules from one location to another, but rather by their simply being passed on by carrier compounds as in a bucket brigade. Energy is preserved and the precision of chemical conversion ensured by natural selection of the survivors. Functional complexity grows inevitably from the failure of self-maintenance; this failure is equivalent to death.

Non-equilibrium thermodynamic systems driven by energy and material flow (mediated by organophosphate compounds) have been life's mode of growth since its inception (Schneider and Sagan 2005). Carbon dioxide and nitrogen reduced by hydrogen, NH_4^+ in solutions or NH_3 gas, etc. are energetically favored to undergo condensation reactions. These reactions are catalyzed by metal ions so that sugars, amino acids, and fatty acids are synthesized. After membrane enclosure of these reactions, others were generated through activation of high-energy compounds. The watery membrane-bounded system became and remained dynamic since the moment of viability.

Contrary to popular scientific wisdom, no flagrant discontinuity distinguishes pre-cellular from cellular dynamics. Growth that requires energy and material flow generated metabolic autocatalytic cycles that led to integration of the genetic system in what became the prokaryotic cell. Cells retained the metabolic systems that preceded them. The minimal prokaryotic cell, with its coupled metabolic-genetic system and high heritability, reproduced beyond the capacity of its environmental support; it was, therefore, subjected to and made more efficient by incessant natural selection.

Life, belonging to a sustained growth process, continues incessantly to grow. Without energy flow, irreparable losses occur. Life forms, past, present, and future, are stabilized dynamic growing systems cellularly separated by membrane boundaries.

Even if, as is likely, my detailed scenario contains errors, the idea that life is an extremely low probability singularity event is an erroneous inference from the "cake mix" (building block/assembly) model. Life's origin, metabolism, and evolution are parts of a single chemical process of energy-driven growth. All cells and organisms made of cells are "selves." This chemical process from growth to life probably appeared in the lower Archean eon (more than 3,400 million years ago) in an astonishingly short time.

3

Alarmones

Antonio Lazcano, Arturo Becerra, and Luis Delaye

Some of the earliest chemicals to arise in the biochemical growth process described by Day in the preceding chapter appear to be intimately connected with sensitivity to the world outside cells. This chapter details the remarkable conservation of cyclic adenosine monophosphate (cAMP), a social hormone related to sensation and action. Lazcano, Becerra, and Delaye locate the roots of sensation in an ancient family of chemicals called "alarmones"—small ribonucleotide-derivatives that are activated when cells sense thirst, starvation, or other environmental threats.

Living beings do not live in blissful isolation. They incessantly exchange matter and energy with other organisms and with their surroundings, in which a wide array of sensory and signal chemical systems intervene. Comparisons of organisms' genomes suggests that a number of the components found in plant and animal sensory systems, including cell-to-cell signal transmission, originated in our bacterial ancestors.

Bacteria traditionally have been depicted as passive "mere infectious agents" that lack an ability to perceive their surroundings. This is a human prejudice. In view of their vast metabolic diversity and broad geographical distribution, their perception may surpass our own. Prokaryotes possess elaborate sensory systems that allow complex monitoring of and diverse responses to their environment. Long ago, early life forms evolved receptors that sensed their environment, taking cues from intracellular and extracellular chemical signals. Transcription of the genetic code to make RNA and protein depending solely on the sensing of small molecules by RNA, and metabolite-responsive riboswitches (i.e., small signal RNAs), suggest that RNA functioned to sense metabolites before proteins evolved (cf. Winkler et al. 2004).

Stress conditions are sensed by alarmones (*alarm* + hor*mones*), small signal metabolites that are synthesized rapidly when intracellular

starvation occurs (figure 3.1). Alarmones are relatively simple compounds made by nucleotide-modifying biosynthetic pathways mediated
by enzymes that may have appeared in RNA/protein cells before DNA
genomes evolved. Together with other ribonucleotide derivatives, these
compounds sense and control multi-enzyme pathways of metabolism,
that is, the metabolic routes including the biosynthesis of purines, vitamins, and amino acids. The biological distribution of adenylyl cyclase
(AC) in the three major biological lineages Bacteria (Prokaryotes,
subkingdom = Domain Eubacteria), Archaea (subkingdom = Domain
Archaea), and Eukarya (superkingdom = Eukarya) suggest that cAMP-
synthesizing enzymes were probably present in the last common ancestor
of all extant life forms. (For the authoritative and consistent classification
of life on Earth used in this book and summarized in appendix A, see
Margulis and Chapman 2010.) Both direct biochemical search and computer genomic analysis have failed to identify other alarmones (such as
ppGpp and pppGpp) and their biosynthetic enzymes in archaea (=
archaebacteria), although they are present in bacteria (= eubacteria) and
eukaryotes. Bacteria appear to have explored the widest range of alarmone biosyntheses and utilization. Various alarmones help sense lack of
food (amino acids or sugars), environmental insults (such as elevated
ambient temperatures), the presence of ethanol and other alcohols, excess
salt, and many oxidants. All known alarmones are modified purine-
ribotides. They include cyclic AMP (cAMP, adenosine 3',5'-cyclic monophosphate), cGMP (guanosine 3',5'-cyclic monophosphate), AppppA
(diadenosine tetraphosphate), and ZTP (5-amino-4-imidazole carbox-
amide riboside 5'-triphosphate). All alarmones so far are related to
purine and none to pyrimidine metabolism. Guanosine tetraphosphate
and guanosine pentaphosphate are also alarmones. They inhibit stable
RNA synthesis in bacteria. The chemistry and the broad biological distribution of purine alarmones such as cAMP suggest that they evolved
as a stress-sensing regulatory system in RNA/protein cells even before
the evolution of DNA genomes.

Early Evolution: Life without DNA?

Because of the small size and chemical nature (figure 3.2) of these chemicals, alarmones cannot be used to construct phylogenies. Nonetheless,
their evolutionary history can be reconstructed by comparing the distribution and phylogeny of the enzymes involved in their biosyntheses.
Lateral gene transfer hinders the reconstruction of early biological

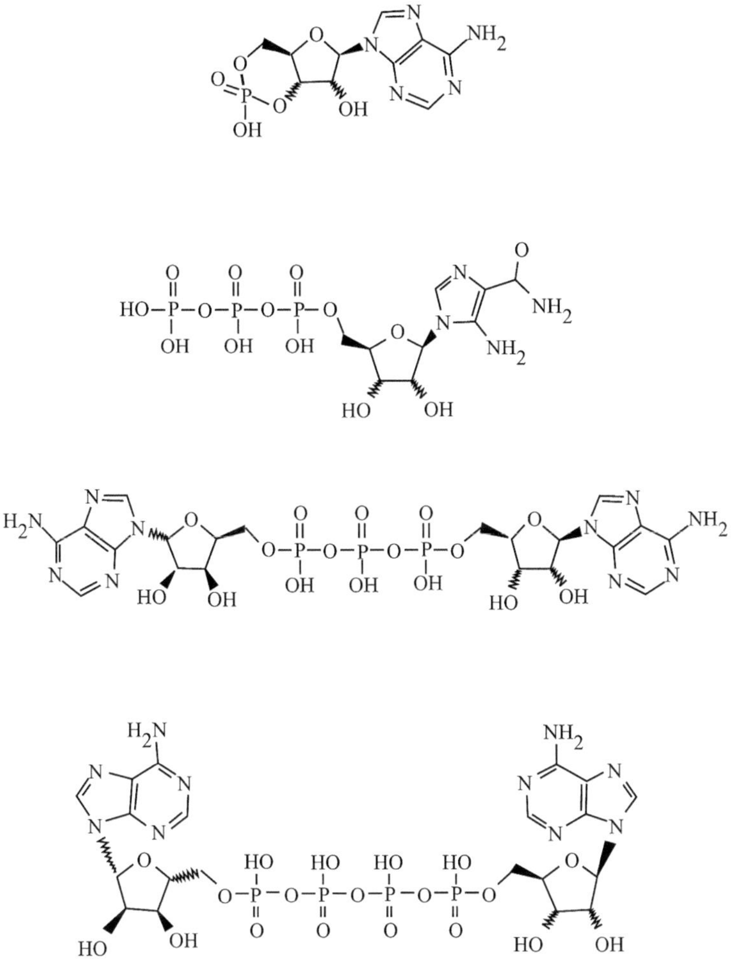

cAMP

ZTP

ApppA

AppppA

Figure 3.1
Four alarmones, examples of structures related to RNA purine nucleotides.

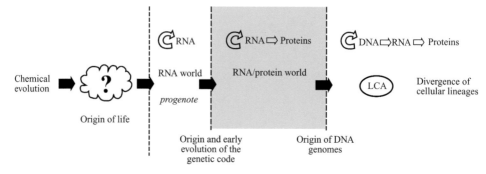

Figure 3.2
Hypothesized steps in the earliest evolution from non-life to life's last common ancestor (LCA). "C-shaped" arrows indicate self-replicating molecules.

evolution, and sequences involved in alarmone biosynthesis have not escaped this fate. However, evidence suggests that the "last common ancestor" (LCA) of all extant life forms—a modern type of prokaryote that synthesized, released, and responded to alarmones—was already endowed with elaborate sensory systems (Becerra et al. 2007). When and how did these molecular sensory and regulatory systems first evolve?

Although how life originated is not known, the discovery of catalytically active RNA molecules gives credibility to the hypothesis that during early stages of biological evolution replication and catalysis were largely mediated by RNA molecules. A stage called "the RNA world" hypothetically existed during the early Archean eon (Joyce 2002). The fact that RNA molecules by themselves perform all the reactions involved in peptide-bond formation suggests that protein biosynthesis evolved in this RNA world. Four of the central reactions involved in protein biosynthesis are catalyzed by ribozymes whose complementary nature suggests that they first appeared in the RNA world, as summarized by Kumar and Yarus (2001). This, in turn, may have led to an RNA/protein world that preceded modern-type cells in which DNA is the genetic polymer. We suggest that alarmones may have first appeared during this early evolutionary stage.

Early Evolution of cAMP and other Alarmones

Cyclic AMP appears to be the most widely distributed alarmone (table 3.1). Since the intramolecular cyclization reaction required for cAMP from AMP (a mononucleotide present in RNA) is equivalent to polynucleotide elongation, it could have been catalyzed in an RNA world devoid of proteins. In any case, cAMP may have already been present in the last common ancestor (LCA). The ubiquity of cAMP probably reflects both its antiquity and the polyphyletic origin of the adenylyl cyclase (AC) enzymes involved in its biosynthesis (Danchin 1993). Structural analyses of ACs have led to their classification into six different families of proteins, of which the AC2 family may be the most ancient. The wide distribution of sequences and catalytic domains involved in its biosynthesis (Aravind and Koonin 1999), which may have been recruited from enzymes involved in nucleic acid replication, is consistent with an early emergence of cAMP.

Although alarmones are found in all three primary biological lineages or domains—Bacteria, Archaea, and Eukarya—their highest structural and functional diversity is found in Bacteria. Biosynthesis of guanosine 5'-diphosphate 3'-diphosphate (ppGpp, or guanosine tetraphosphate) and guanosine 5'-triphosphate 3'-diphosphate (pppGpp, or guanosine pentaphosphate), which act as general signals for amino acid starvation conditions by adjusting the rates of the synthesis of basic components of protein synthesis, is mediated by GTP pyrophosphokinase, which catalyzes the transfer of phosphates. Somewhat equivalent chemical processes are involved in the biosynthesis of other alarmones, including AppppA and ZTP (figure 3.3).

Conclusions

Analysis of the basic traits of living systems supports the hypothesis that RNA molecules played a major role in reproduction and metabolic pathways during early stages of biological evolution. Components of the RNAs essential to the metabolism and reproduction of cells as the units of life also emerged early as simple sensors (figure 3.4). The evolution of RNA-based controls helps to explain the continued existence of RNA riboswitches in protein synthesis. It seems likely that cAMP and other alarmones are "metabolic fossils" of a stage in the evolution of chemical signaling and metabolite sensing that existed in RNA cells more than 3,500 million years ago.

The presence of cAMP and the ubiquitous distribution of AC2s as defined by the comparison of sequences of ribosomal RNA and other molecular markers, suggest that cAMP was already present

Table 3.1
Distribution of alarmone-synthesizing enzymes.

Alarmone	Enzyme	Organism in which first characterized	Phylogenetic distribution		
			Eubacteria	Archaebacteria	Eukarya
cAMP	AC[1] class I (enterobacteria)	Escherichia coli	Gamma-proteobacteria		
cAMP	AC class II (calmodulin-activated)	Bacillus anthracis	Proteobacteria, firmicutes		
cAMP	AC class III (palm domain)	Rattus norvegicus	Several eubacteria[2]	Few euryarchaea[3] (methanogens)	Several eukarya[4]
	AC class IV (AC2)	Aeromonas hydrophila	Several eubacteria[5]	Several archaebacteria[6]	Several eukarya[7]
	AC class V	Prevotella ruminicola	P. ruminicola (bacteroides)		
	AC class VI	Rhizobium etli	Alpha-proteobacteria, spirochetes		
ZTP	PRPP synthetase (5-phosphoribosyl-1-PPsynthetase)	Homo sapiens	Many eubacteria	Many archaebacteria	Many eukarya
ppGpp	GTP pyrophosphokinase	Salmonella sp.	Several eubacteria[8]	—	Anopheles gambiae, red algae, plants[9]

Table 3.1 (continued)

Alarmone	Enzyme	Organism in which first characterized	Phylogenetic distribution			
			Eubacteria	Archaebacteria	Eukarya
ApppppA	ATP adenylyltransferase	*Sacharomyces cerevisiae* (yeast)	Cyanobacteria	—	Ascomycotous fungi; red algae, slime molds[10]
AppppA	Diadenosine 5, 5-P1, P3-triphosphatase	*Sacharomyces cerevisiae* (yeast)	Most eubacteria	Most archaeabacteria	Most eukarya

1. AC: adenylate cyclases.
2. Cyanobacteria, actinobacteria, proteobacteria, planctomycetes, spirochaetes. Not present in *Bacillus/Clostridium, Deinococcus,* chlamidiae, gamma and alpha proteobacteria (Shenoy and Visweswariah 2004).
3. Present in *Methanosarcina acetivorans* str. C2A, *Methanopyrus kandlery*, and *Methanothermobacter thermoautotrophicus* (Shenoy and Visweswariah 2004).
4. Lacking in *Arabidopsis thaliana.*
5. Cyanobacteria, actinobacteria, bacteroidetes, proteobacteria, planctomycetes, spirochaetes, firmicutes, bacillariales.
6. AC2 is not present in the archaeabacteria: *Thermoplasma acidophilum, T. volcanium,* or *Picrophilus torridus.*
7. Metazoa, mycetozoa, diplomonadida, plantae. (Lacking in fungi.)
8. Aquificae, actinobacteria, *Deinococcus-Thermus,* bacteroidetes, proteobacteria, spirochaetes, firmicutes, cyanobacteria, thermotogae, fusobacteria, *Chlorobium.*
9. *Anopheles gambiae,* Diptera (*Anopheles*).
10. Protocists: the red alga is *Cyanidioschyzon merolae*; the cellular slime mold is *Dictyostelium discoideum.*

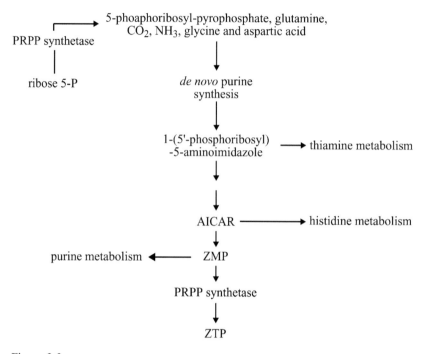

Figure 3.3
The precursors to the earliest alarmones are common universal metabolites such as phosphorylated ribose, carbon dioxide, ammonia, glutamine, and aspartic acids. From this we infer that the alarmone pathway is ancient and embedded in early cells.

in the LCA. Cyclic AMP has been conserved across prokaryotes and nucleated cells. From the fact that multiple biosynthetic enzymes of cAMP metabolism have been identified we infer that cyclic AMP was selected as a secondary messenger by manifold convergent processes. The phylogenetic distribution of adenyl cyclases defies simple explanation. The genes involved in cAMP synthesis may have undergone branch-specific gene duplications, lateral transfer events, and secondary losses, and may have evolved independently in different prokaryotes. Such diversity and unequal distribution not only indicate the need for more detailed studies of alarmone evolution, but also point to the wide diversity of sensing strategies and signal mechanisms (in later pathways such as steroids, ethylene, and lectins) that the biosphere has successfully used during over 3,000 million years of incessant change.

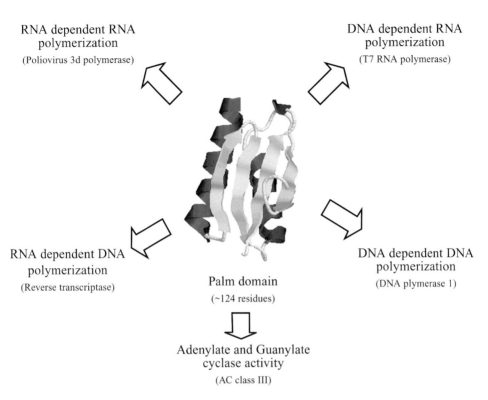

RNA dependent RNA
polymerization
(Poliovirus 3d polymerase)

DNA dependent RNA
polymerization
(T7 RNA polymerase)

RNA dependent DNA
polymerization
(Reverse transcriptase)

Palm domain
(~124 residues)

DNA dependent DNA
polymerization
(DNA plymerase 1)

Adenylate and Guanylate
cyclase activity
(AC class III)

Figure 3.4
DNA is synthesized from RNA, a fact that support our hypothesis that RNA evolved first.

Acknowledgments

Work reported here was supported by project DGPA IN-111003 (UNAM, Mexico). Part of this work was completed during a leave of absence at Rice University, where AB enjoyed the hospitality of Dr. Janet Seifert. LD thanks the 2000 Planetary Biology Internship NASA program for its support for a training visit at the laboratory of Dr. Peter Gogarten.

4

Early Sensibilities

Kenneth H. Nealson

What do bacteria sense, how do they sense, and how does their sensing affect them as individuals and as communities? Here a microbiologist asks "What is the microbial IQ?" and "How do bacteria use their metabolic gifts to move, to detect light and oxygen, and, in general, to survive?" Nealson surveys what we know about bacterial sensation and points out challenges and research opportunities related to the origins of microbial sensitivity.

The ability to sense the environment and to respond to it by measurable movement is the essence of behavior. There is a general sense that the responses of life forms became more complex with time. But what do organisms sense? How and why do they sense their surroundings? How do they respond to their sensations? Without attempting to be encyclopedic, I attempt here to stimulate thought about the earliest chemistry- and gene-modulated appearance of sensations in our cellular ancestors.

The smallest, morphologically simplest cells, the bacteria in the broad sense, were unified under the term "prokaryote" in the early 1960s. Since then, "prokaryote" has been used to denote life composed of bacterial rather than nucleated cells. Prokaryotes do not enclose their genes, their DNA, within a pore-studded nuclear membrane. The word "prokaryote" has been challenged recently because it includes two distinct groups of microorganisms: the Bacteria (also known as eubacteria or true bacteria) and the Archaea (also called archaebacteria) (Woese et al. 1990). Part of the rationale for rejection of the term "prokaryote" is that some bacteria taxa (e.g., *Gemmata obscuriglobis*, a member of the group Planctomycetales) have membranous structures that surround their DNA (nucleoids). The features of true nucleated cells include many other characteristics, such as incessant intracellular movement (e.g., cytoskeletal activity mitosis, chromosome movement and that of other organelles, phagocytosis, and exocytosis). Also, the physical separation of translation (messenger RNA synthesis) and transcription (protein synthesis) is not found in either prokaryotes, either bacterial or archaeal.

Carl Woese admonishes that we now let "prokaryote" mean only cells that are not eukaryotic, with no monophyly implied. He believes that "prokaryote can still be used as long as conveniently needed, but it will now imply nothing about relationships or structure." I think we need either a term to replace "prokaryote"—a term that allows discussion of both archaeal and bacterial cells (which share many properties and which populate ecosystems that may be devoid of eukaryotes)—or a new acceptable term or definition. In the absence of a new word, I abide by Woese's suggestion. I simply define prokaryotes as organisms that are not eukaryotic, because of features shared by bacteria and archaea but not eukaryotes. (See appendix A.) Features shared by the bacteria and the archaea relate nearly exclusively to physiology and ecology. Prokaryotes differ most dramatically in cell structure, details of multicellularity, and behavior. The only definitional difference without exception is the presence in all eukaryotes at all times of the double-membrane-bounded nucleus that encloses protein-studded chromosomes segregated to offspring cells by a cytoskeleton, the mitotic microtubular apparatus.

Several of the properties unique to prokaryotes may be relevant to the nature of the sensory self. Prokaryotes, in general, are small. They are not capable of phagocytosis or mitosis. Thus, their metabolism is restricted to chemistry-driven lifestyles, dominated by small cells (high surface-to-volume ratios), extracellular enzymes, transport systems, and diverse metabolic abilities. Their chemical signals allow them to identify energy sources (i.e., electron donors and acceptors, light) rather than other organisms. Many of the properties that characterize prokaryotes are related to their metabolic versatility. Their modes of metabolism include chemolithotrophy, anaerobic respiration, methanogenesis, nitrogen fixation, and photosynthesis. Eukaryotes, by contrast, show limited metabolic features: nearly all have aerobic respiration, have some form of heterotrophy (in the dark, at least), have predatory and other complex behavior, and synthesize secondary metabolites. Further, animals and plants show complex tissue-level morphology and meiotic sexuality. Embryos are among the many features limited to animals and plants.

Self

This book is deeply concerned with defining the sensing, ultimately thinking self. An investigation of the self begins with an investigation of the cell. In the structurally and organizationally complex eukaryotes, the cell is the unit of life. An organism is made of one or more cells, and often of a variety of cells in tissues. Each cell contains the same genetic information but expresses it differently. However, all eukaryotes, whether animal, fungal, plant, or protoctist, live in incessant interaction with the bacterial world. The many bacterial symbionts in physiological, biochemical, and genetic communication, those intertwined with the eukaryotic self, must be considered. Thus, to define self is a challenge for any eukaryote.

Is it easier to define self for the prokaryotes? Again, the cell as the basic unit of life is often considered to be the unit of self. Great interest in the study of individual microbes is thwarted by methodological limitation. A population of bacteria through their physiological responses displays different cell states. Some grow rapidly, some slowly, and some not at all. Even if cells in the population have the same genetic endowment, the same set of genes, they differ. A biofilm community with rapidly growing cells on its periphery or its exterior will have quiescent cells in the interior and nearly non-functional cells at the base. Each cell is distinct from the others. Taken together, they form the basis for activity of the whole. In fact, the robustness and the longevity of the whole depend on the components. Bacteria grown in controlled conditions in a chemostat help us understand their potential, but nature is hardly ever so defined. Physiological "differentiation" is probably a rule of microbial survival. Sensing when to stop growing may be as important to the long-term survival of the prokaryotic self as is rapid growth.

The collection of similar cells that interact as a population may define self. This idea is as incomplete as a eukaryotic self without its bacterial partnerships. In nature, eukaryotes and their bacteria (whether eubacteria or archaeabacteria) are seldom if ever found alone, as "individuals." Although strains and species vary, the physiological types in nearly any ecosystem are quite predictable from an energetic perspective. Although the prokaryotes are remarkably diverse biochemically, few are generalists. All, whether clumps of a few hundred or masses of millions, require a continuous flow of energy as oxidizable substances (e.g. methane, hydrogen, sulfide) or light. All need nutrients. They must build usable forms of carbon and nitrogen compounds into their bodies. In contrast to populations of the same kinds of microbes, natural microbial communities operate through complex interactions of energy and matter flow. The members of a community harvest environmental requirements that serve to maintain function. The continuation of survival, growth, and, eventually, reproduction mandates the continuous flow of energy and matter through the community.

Despite the difficulties of defining self, sensitivity and response to details of the environment are as crucial to the life of prokaryotic communities as to the eukaryotic world.

What Stimulates Bacteria?

Prokaryotes respond to a surprisingly large set of environmental variables and to changes in them: light, temperature, salinity, availability of energy sources, electron acceptors, oxygen and other gas composition, magnetic fields, specific informational molecules, such as alarmones (chapter 2), that act as hormones or signals. Prokaryotes are rather constrained in these responses relative to their eukaryotic counterparts. Animals, for example, have sensors for smell, water, salt, temperature, taste, pain, and sound. Prokaryotic responses tend to be less localized and more direct. Bacteria rapidly move toward or away from stimuli. The turning on or off of specific genes in response to stimuli is often the way sensitivity is detected. The prokaryotes as individual cells are almost certainly constantly sampling and responding to environmental variables and their changes. This is the nature of vigilance (sensing and response) that we want to reveal.

Chemotaxis, the movement of prokaryotes up or down a chemical gradient, is commonly measured. Both positive (attraction) and negative (repulsion) responses are known, and, while a number of approaches are used to detect the signal and elicit the response, in general the systems involve two components: a sensing device that contains the sensor and response regulator, and a motility mechanism (flagellar or gliding motility) that allows the response to be carried out. Most studies are done in liquid media with motile flagellated bacteria, but clearly the same types of responses occur on surfaces or in semi-solid media. Of great importance to activities of microbes is bacterial sensitivity to immediate surroundings. Interactions lead to the development of communities we detect as microbial mats and biofilms. Well-demonstrated examples are the gliding bacteria (myxobacteria), which use chemotactic signals to find their prey and swarm toward it. Other chemical signals initiate differentiation to make "bacterial trees" with myxospores (Shapiro and Dworkin 1997).

Components of signal sensation appear to be conserved. "Genomic" methods permit identification of genes, estimation of the phyletic distribution of the tactic responses, and variations on the chemical theme. The best-known system for tactic response may be that of the so-called energy taxis of *Escherichia coli* (Alexandre et al. 2004). These bacteria synthesize sensor molecules called methyl-accepting chemotaxis sensory proteins (MCSPs), which are used to sense a concentration gradient. They detect an energy source or a toxin and respond by swimming toward or away from the stimulus. If a concentration increases, methylation of a sensor specific for that concentration occurs. As long as it continues to increase, flagellated bacteria continue to swim toward the highest concentration. If the concentration of the sensor molecules decreases, MCSP methylation ceases. The methylated proteins become non-methylated. This change leads to a reversal in the direction of flagellar rotation. In *E. coli* the reversal leads to tumbling activity that induces the swimmer to change direction. The flagellar direction reversibly reverses. The concentration change, the methylation-demethylation, allows the bacteria to accomplish a directed random walk. The bacteria always grow, glide, or swim toward the stimulus of interest or, alternatively, away from toxins. Some bacteria have a single polar flagellum at the end of the bacterial body. When reversal occurs, they simply swim in the opposite direction, i.e., the flagellum pulls rather than pushes the cell. A similar biased random walk results; however, as the reversal is seldom perfect, each reversal results in a slightly different swimming angle (Armitage 1999).

In *E. coli* this behavior is understood. A series of transmembrane chemoreceptors, the methyl-accepting chemotaxis sensory proteins (MCSPs), receive an environmental signal. They interact with the cell's interior by means of a docking protein, **CheW**. Interaction of MCSP with **CheW** leads **CheA**, a cytoplasmic histidine kinase enzyme that adds a phosphate group to a protein amino acid, to become autophosphorylated. Phosphorylated **CheA** then transfers its phosphate to **CheY**, which, in the phosphorylated state, binds directly to a switch in the flagellar motor. The activated switch changes the direction of rotation of the flagellar bundle. My simple description omits a number of details that make regulation of bacterial chemosensitivity subtler than described here.

The preponderance of similar genes across a wide range of bacteria leads us (Alexandre et al. 2004) to suspect that similar sensory systems exist among other prokaryotes. However, most details of chemotaxis have yet to be studied. Many variations on this general theme probably will be discovered.

Energy Taxis

In addition to chemotaxis, prokaryotes exhibit phototaxis to kinds and levels of light, and aerotaxis to levels of oxygen (Adler 1988). This, along with reports of movement in response to other alternate electron acceptors (Taylor et al. 1979; Nealson et al. 1995) and of positioning in redox gradients (Grishanin et al. 1991), leads us to think that these responses may be conceptually linked by "energy taxis." Instead of responding to environmental gradients of specific chemicals, each of which requires its own sensor protein, the cell may constantly monitor its internal energy state, and exert control over the flagella motility in response to a combination of the intracellular redox level (oxygen concentration) and proton motive force (hydrogen ion concentration, acidity).

Aer, a transducing protein involved with aerotaxis, may be involved. Aer contains a flavin-adenine dinucleotide bound to a protein component that probably changes configuration as a small part of it is oxidized or reduced (Alexandre et al. 2004). Many different energy inputs, including light, inorganic energy sources, organic energy sources, and electron acceptors, may elicit a common response. One expects that many transducers (energy-conversion molecules) must have evolved to accomplish such complex regulation. Prokaryotes have metabolism that requires monitoring for the bacterium to locate itself in the "right place." Photosynthetic bacteria capable of heterotrophic metabolism come to mind. They orient to light intensities and qualities, and to concentrations of sugars and other foods.

Multiple Systems for Sensing

Genomics and the full sequencing of more and more microbial genomes have enabled us to look at genes such as those that determine methyl-accepting chemotaxis sensory proteins. Virtually all prokaryotic genomes reveal the presence of genes for MCSPs, presumably for tactic (mechanosensitivity) abilities. Large differences exist among the prokaryotes with regard to the number of MCSPs present. Members of the proteobacteria groups often have 30 or more, whereas methanogens and other archaea have no more than 10 (Alexandre et al. 2004).

Shewanella oneidensis MR-1, a metal-oxide-reducing gamma proteobacterium, is closely related to *E. coli* (Myers and Nealson 1988). Because it synthesizes more than 25 MCSPs, this strain of *Shewanella* might be expected to be strongly chemotactic in comparison to *E. coli*. However, we (Nealson et al. 1995) have shown that it is not attracted to any energy sources, whereas *E. coli* is chemotactic to a wide range of different energy sources. Rather, *Shewanella* responds to gradients of electron acceptors it uses for respiration.

Multiple copies of the receptors and the transducers signal a complex and changing lifestyle, whereas a single copy denotes an organism with limited niche versatility. *Shewanella oneidensis* is found in environments that show frequent and rapid changes in oxygen concentration. Is this general? Given the distribution and the abundance of these genes, can we infer properties of the environment from metagenomic analyses?

The term "microbial IQ," defined in part by the number of sensory systems found in the genome of the organism in question, is used to describe the ability of a prokaryote to sense and respond to changing environments (Galperin and Gomelsky 2005). Not surprisingly, bacteria in extremely stable non-fluctuating environments are on the lower end of the "microbial IQ" curve. Apparently, they need very little response to their predictably stable environment. In contrast, microbes constantly challenged by dramatic environmental change seem to be quite "intelligent" indeed!

Cell-Cell Communication and Community Interactions: Quorum Sensing

Some bacteria synthesize compounds that specifically induce processes such as "turning on" their internal lights—their bioluminescence. Only when the population becomes sufficiently dense does a certain molecule (the autoinducer, acyl-homoserine lactone) accumulate in the surroundings at a concentration high enough to induce the synthesis. This "autoinduction" was once thought to be restricted to marine bacteria. Now we know it is widespread. Called "quorum sensing," such population responses regulate a wide range of activities in symbioses that include pathogenesis (Dunny and Winans 1999). When enough bacteria of this kind are concentrated in the same place, the population produces a toxin or a signal molecule.

Some bacterial taxa delay the expression of a certain gene until the population density of that group reaches a minimal level. The original notion that a single molecular type is involved in quorum sensing has been superseded. Multiple autoinducers are found in some species of bacteria (Henke and Bassler 2004). The genes involved, denoted "lux genes" because of their part in the luminescence system, in fact have wide-ranging functions not limited to light sensitivity (Engebrecht et al. 1983).

Peptide inducers, small protein-like molecules, appear to function in spore-forming *Bacillus* strains (Gram-positive bacteria) that turn on their spore-making processes in a way that is similar to the induction of bioluminescence ("cold light") in luminous Gram-negative bacteria.

The Great Unknown

Whereas it is easy to see how the ability to sense and respond to the environment on both short and long time scales is of great use to swimming prokaryotes, we have only a glimpse at the way complex signals and responses are integrated ecologically in nature. We have few examples of the details of sensory capabilities and responses of bacteria to mechanical stimuli, gravity, light, heat and cold, Earth's magnetic field, salt and water concentrations, alarmones, and many other simultaneously and continuously changing environmental stimuli. Even in bacteria, it is not yet possible to define a prokaryotic self. So great are the unknown ways in which populations of bacteria respond to members of their living communities and to cyclically changing physico-chemical signals that I can only admire the daring of my co-authors (see chapters 7, 15, 19, 20, and 23) in attempting to answer questions of the self in humans and other primates.

II

Groups

5

Smart Bacteria

Eshel Ben-Jacob, Yoash Shapira, and Alfred I. Tauber

Here Ben-Jacob, Shapira, and Tauber explore how bacteria together sense the environment, extracting matter and energy from it as they engage in de facto *cognitive processes.*

Eons before humans, bacteria inhabited a very different Earth. As the earliest life form, they countered spontaneously increasing entropy. They converted high-entropy inorganic substances into low-entropy organic molecules (Ben-Jacob et al. 2004, 2006). They paved the way for other life forms by changing harsh physical and chemical conditions on Earth's surface and its atmosphere into a life-sustaining environment. They enriched the atmosphere with oxygen and loaded water and soil with minerals and organic nutrients, enabling their descendants to flourish. Bacteria are, simply, indispensable to all other life on Earth.

Bacteria are not simple, solitary creatures of limited abilities, as was long believed. Indeed, a fundamental aspect of bacteria is their cognitive ability. The idea that bacteria are simple and solitary was perpetrated over years of their growth in artificial conditions in the laboratory. They all require one or more of the three major forms of energy: light, organic, and inorganic chemical oxidation. Their habitats range from sulfuric hot springs to icebergs. Under the demands of the wild, these versatile life forms work in teams. They live in dynamically stable communities that require active communication (Shapiro and Dworkin 1997; Ben-Jacob et al. 2004). Bacteria are "smart" in their use of cooperative behaviors that enable them to collectively sense the environment. They use advanced communication, and they lead complex social lives in colonies whose populations exceed the number of people on Earth (Ben-Jacob et al. 2004).

To deal with changing environmental hazards, bacteria resort to a wide range of cooperative strategies. They form complex patterns so as to function efficiently. They modify their colonial organization (for example, they alter the spatial organization of the colony in the presence of antibiotics) to optimize their survival. Bacteria, we argue, have collective memory by which they track previous encounters with antibiotics. The colony of individuals, the social group, gleans information from the environment. They "talk" with one another, distribute tasks, and convert their collective into a huge "brain" that processes information, learns from past experience, and, we suspect, creates new genes to better cope with novel challenges (Ben-Jacob et al. 2004).

Schrödinger's (1994) "consumption of negative entropy" requirement offers a way to understand these complex phenomena. The energy, matter, and thermodynamic imbalances provided by the environment led Schrödinger to propose that life required consumption of negative entropy, i.e., use of thermodynamic imbalances in the environment. When we adopt his perspective of thermodynamics, each bacterium becomes a hybridization of an "engine" that uses imbalances in the environment to do work, and a "machine" that uses this energy to act against the natural course of entropy increase, for the synthesis of organic substances. We propose a third information-processing system for the coordination and synchronization of the engine and the machine. A living bacterial cell is analogous to a complex artificial cybernetic system or a "chimera" composed of information-processing systems and at least two thermodynamic elements. Their outer membranes enable them to sense the environment and to exchange energy, matter, and information with it. In conjunction with the surrounding conditions, the cell's internal state and stored information regulate the membranes.

In addition to "consumption of negative entropy," organisms sense the environment so as to extract latent embedded information (Ben-Jacob et al. 2006). By "latent information" we mean that information embedded in the environment, once processed cognitively, initiates change in the sensing organism. Information induces change; hence it is used to generate an internal condensed description of the environment.

An individual bacterium can sense only a limited volume of its aquatic habitat before it reproduces, but a colony composed of thousands of millions of bacteria is capable of sensing large volumes of habitat, and over long time periods. For such coherence, bacterial colonies evolved communication abilities to exchange information about myriad detections stored as newly acquired information. As a member of a complex superorganismic colony, each bacterial unit (cell) possesses the ability to sense and communicate with the others. Together they constitute a coordinated collective that performs integrated tasks in communication with the behavior of others. Collective sensing and cooperativity are intrinsic to microbial communication. Multicellular superorganisms (communities) generate in their constitutive elements (individual bacteria) new traits and behaviors not explicitly stored in the genes of the individuals. Therefore, bacteria do not store genes in their cells that code for all the relevant information required for the patterns made by the colonies. Contextual information is generated by use of both genetically stored information on how to produce perceptive faculties and information extracted from the environment. The bacteria use their intracellular flexibility (e.g., signal transduction networks, gene signaling, and genomic plasticity) to create the patterned colony. They maintain integrity and generate morphological change by sharing interpretations of chemical cues. The dialogs produce meaning-based communication (Ben-Jacob et al. 2004) that permits purposeful alteration of colony structure and group decision making. These bacterial faculties represent the origins of cognition. Although bacteria lack mammalian organ-system based communication capability, they display precursors to such more complex functions.

Systems: From Thermodynamic to Cybernetic

Bacteria evolved a plethora of different modes of energy using chemical gradients (entropy imbalances) that convert high-entropy inorganic compounds such as atmospheric CO_2 into low-entropy life-sustaining organic molecules such as organic acids, including amino acid, precursors to proteins, and ribonucleic acid. Our scientific advances help reduce the mystery of the origin of chemoautotrophy. (See chapter 4.) Every living organism and each individual cell in multicellular organisms, we suggest, is an information-based cybernetic cognitive chemical system that meaningfully functions to ensure the flow of energy and matter of autopoiesis.

We propose that biotic systems are analogous to chimeras of three types of artificial machines: thermodynamic engines, pumps, and information-processing systems. Bacteria are analogous to complex human-made cybernetic systems composed of information-processing and thermodynamic machines, ones that reduce entropy by use of environmental energy.

An individual bacterium is not comparable to a single man-made machine, but rather to an entire factory composed of many interacting artificial machines and information processing systems that by the exchange of energy and matter generate new information.

Cooperative Behavior in Spore-Forming Myxobacteria

Collective sensing is illustrated by bacterial behaviors. In *Myxococcus xanthus*, foraging parties of cells comprise the advancing edge of the colony. Upon chemical sensing of a food source, scouts return environmental information to the entire colony, which expands by gliding movement and cell division growth toward the newly detected food source (Shapiro and Dworkin 1997).

Bacterial "discourse" is illustrated in the starvation response of sporulation. When growth is stressed by desiccation or starvation, members of the colony transform into inert desiccation-resistant propagules, in this case spores. Sporulation begins only after "consultation." A collective perception of stress occurs as starved cells emit chemical messages to communicate a message for or against sporulation. Once each cell in the collection has sent its preference (chemical message) and responded to the others, sporulation is initiated if the "majority is in favor."

Communication-Based Self-Engineered Organization

Social lives of bacteria are revealed by exposure to adverse growth conditions that mimic those encountered in nature. The patterns (figure 5.1) are generated in response to growth on nutrient-poor, hard surfaces. The colony produces a lubricating fluid layer that permits swimming on hard surfaces to cope. The individual bacteria at the anterior of the moving colony push the layer forward to pave the way for colony expansion. By adjustment of lubricant viscosity, the bacterial cells remain in contact such that the population density is retained at a high enough value to ensure protection and efficient use of nutrients.

After inoculation onto a softer substrate, a condition more amenable to swimming, the bacteria display radical variations in colony pattern. The huge population of cells seen as a microscopic branch in figure 5.2 exhibits chirality. The branches always curl in the same direction. A genetic change accompanies the new colony structure as each bacterium grows longer. This helps them move with coordinated motion as shown in figure 5.3.

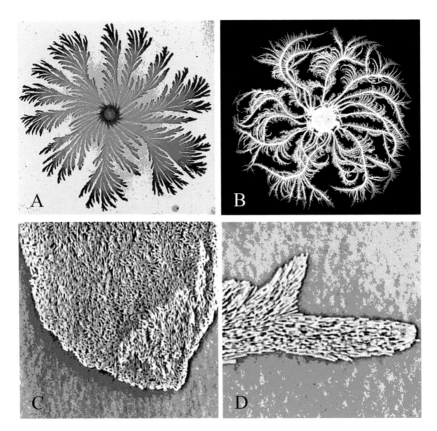

Figure 5.1

Paenibacillus dendritiformis when these bacteria are grown in a Petri dish on hard agar that lacks sufficient food. Far from being shapes of mere aesthetic beauty, these colonial structures reflect the self-engineering skills of bacteria. The spreading patterns help the colony access more of the scarce food in the most efficient way under the given conditions. Panel A shows the ordinary branching pattern, panel B the chiral one (with broken left-right symmetry). The top pictures show the colony patterns. The colonies are a few centimeters in diameter: each contains more bacteria than the number of people at present on the Earth. Panels C and D show the individual bacteria (the small bars) at the branch tips with 500× magnification for panel A and panel B respectively. Video clips of the movement are available on the Internet.

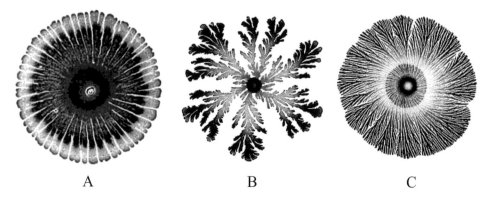

A   B   C

Figure 5.2
Branching growth patterns of *Paenibacillus dendritiformis* bacteria grown in a Petri dish. To self-engineer their colonial structure, these bacteria regulate the balance between attractive and repulsive chemotactic signaling as well as their food chemotaxis. Panel A shows the pattern at higher food levels when attractive chemotactic signaling is activated. Panel B shows the typical pattern when food chemotaxis dominates the growth at intermediate levels of food depletion. Panel C shows the growth at a limiting level of food when repulsive chemotactic signaling is intensified. Note that the pattern is organized into narrow, straight branches. These examples of engineered self-organization enable the colony to make more efficient use of their food while overcoming the challenge posed by the hardness of the agar surface (Ben-Jacob 2003).

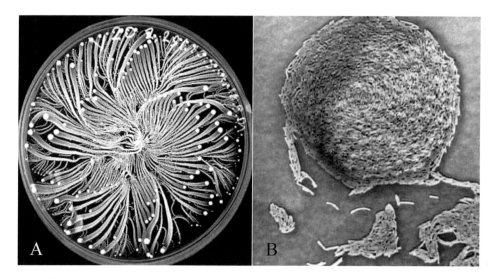

Figure 5.3
Paenibacillus vortex growth on a very hard surface in a Petri dish. In these colonies (panel A), foraging vortices of rotating bacteria grow and move out to cover the hard agar, lubricating the way for their followers. As the colony of *P. vortex* (panel B, ×500) grows by cell division of its constituent prokaryotic cells, it moves and produces a trail of bacteria. They are pushed forward by the bacteria left behind. The process stalls and provides the signal for the generation of a new *P. vortex* clone behind the original one. The original clone leaves "home" the trail as the new entity begins colonization of new territory. Video clips of this movement are available on the Internet.

Chemotactically signaling bacteria inform their peers about the pre-ferred directions for growth. When a rich source of food is detected, attractant chemotactic signals are sent such that peers are called to join the meal. When members of the colony detect regions of low food, strong acids, high salts, low oxygen, or other environmental insults, repellent chemical signals are released that warn other members to stay away.

Learning from Experience

Bacteria may possess epigenetic memory that enables them to keep track of how they handled previous encounters with antibiotics. Colonies cope as if they learned from their past experience by altering their pattern formation upon a second encounter with the same antibiotic (figure 5.4).

Bacterial chemical communication includes assignment of contextual meaning to messages (semantic/syntax functions). For example, during sporulation each cell secretes pheromones to broadcast its own stress and interprets the stress signal received from its peers according to its own present and past conditions. Thus, according to the assessment of its conditions relative to that of its peers, each cell decides to sporulate or

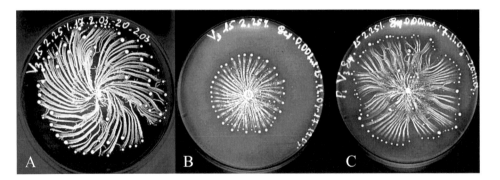

Figure 5.4
Bacterial colonies learn. The response of the *P. vortex* bacteria to non-lethal levels of the antibiotic Septrin. The normal growth pattern in the absence of the antibiotic is shown in panel A. The effect of first exposure is shown in panel B. The response in a second encoun-ter is shown in panel C. The antibiotic stress induces the bacteria to intensify chemotactic attraction and form larger vortices. This clever strategy protects the bacteria since, as in larger vortices, the antibiotic is diluted by lubricating fluid they excrete. The bacteria enhance their repellent chemotactic signaling to push the vortices away from the antibiotic more rapidly. The "higher complexity for better adaptability" behavior is manifested in the fact that the growth pattern in panel B is less complex than that in panel A. Learning from experience is exemplified in panel C. Upon a second encounter with the antibiotic the colony expands faster and the pattern is more complex.

become competent. We interpret this too as a "dialogue" between cells in the colony. Semantics implies that each bacterium possesses some plasticity to assign its own interpretation to a chemical signal according to its own intracellular information state that responds by perception of external conditions. Semantic function implies that no chemical message just triggers a pre-determined pathway, but rather that the environmental chemical signal triggers an intracellular response that involves internal restructuring. The internal response involves the gene-protein synthesis network or even parts of the genome itself.

Bacteria provide a striking example of how organisms change their environment to accommodate themselves, and of how the immediate surroundings should be understood not only as context but as an integratable part of organismic identity. The organism-environment complex co-evolves. In our formulation, cognition and information arise from a conjoined organism-environment construct. This dialectical mode of thought comprehends evolutionary history as a complex interaction of genetic, developmental, and environmental processes that have co-evolved since the early Archean eon.

6

Ancient Architects

Wolfgang E. Krumbein and Celeste A. Asikainen

Congratulating ourselves on our upright stance, our big brains, and our language skills, we consider ourselves the most "evolved" species. But bacteria exceed us in chemical (metabolic) abilities and in importance in distributing the chemical elements in the biosphere. They also build many intricate structures—including, from one way of looking at it, our bodies. Here Krumbein and Asikainen explore another kind of architecture made by collectives of microbes.

Conscious human architecture is considered one of the high points of human civilization. Arches are also built by termites, however, and microbes work manganese and other metals, as well as limestone and other minerals, into functional and beautiful dwellings. Microbial mats from flat, laminated *Microcoleus chthonoplastes*-dominated communities on the shore of the North Sea in northern Germany were grown in the laboratory. They spontaneously generated balloon-shaped protein-lipid exudates 1–5 millimeters in diameter. When these were subjected to ordinary periodic light and temperature alterations with wet-dry cycles, at first they formed balloon-shaped scums that most ambient bacteria were unable to penetrate. These "mini-balloons" floating in the mat mucus or the water column above the laminated mat communities. The mini-balloons were first colonized by one species of filamentous cyanobacteria (*Phormidium hendersonii*). A common diatom, *Navicula perminuta*, glided into the ruptures made by *Phormidium* and then reproduced inside to fill up the balloon. Five of a total of 11 to 13 strains of heterotrophic bacteria were detected inside the balloon. Most of these bacteria have been grown by themselves in pure culture. Their DNA has been sequenced. Spontaneous calcification occurred and ooids (uniform spherical grains of carbonate sand) were released from the ruptured "balloon skins" in approximately half of 350 cases studied. Those that failed to calcify (harden) and release ooids simply disintegrated. These released organic rather than mineralized contents that could not be further studied.

In the second case, a field of concentrically laminated iron-manganese nodules was found at a depth of 10 meters in fresh cold water off the ramp of an abandoned logging camp at the Second Connecticut Lake in Pittsburg, New Hampshire. (See color plate I.) Though the associated bacteria have not yet been identified, the growth pattern from a field of "cow pies" that extend to form a continuous pavement, the presence of a nepheloid biofilm replete with many forms of heterotrophic bacteria, and the laminae suggest that these nodules are a form of rock constructed by microbes, i.e., a microbialite (figure 6.1). In a manner reminiscent of other well-studied stromatolitic communities in the geomicrobiological literature, these sedimentary structures, the intertidal marine oolites, and the submerged freshwater iron manganese nodules form temporary biogenic dwellings for their communities of microbial inhabitants.

Based on our own research and our acquaintance with the geomicrobiology literature, we posit that specific groups of organisms—among

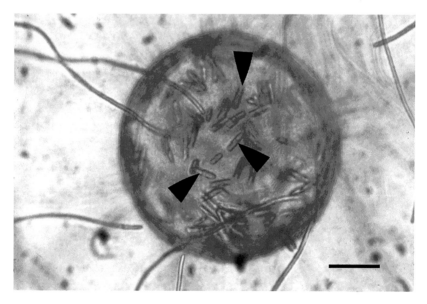

Figure 6.1
Initial steps of organic membrane sphere formation, a precursor to the laminated ooids. Spheres form spontaneously on the surfaces of microbial mats and, in the end, release carbonate ooids, components of oolitic limestone. Oolites, which are fossils, are misclassified as abiotic structures by geologists and ignored by paleontologists. Scale bar = 30 μm. Arrows point to bacterial cells growing by division.

them calcifying cyanobacteria, silicifying diatoms, iron-oxidizing and manganese-oxidizing bacteria, endolithic and epilithic lichen, exopolymeric and rock-eroding fungi, reef-building brachiopods and corals, tower-constructing predatory foraminifera, calcifying alvinid and pogonophoran tubeworms, mound-building termites, tent caterpillars, ground-nesting bees (Turner 2000, 2007)—produce highly patterned dwellings. Complex community-built structures are known from the fossil record to date to long before the evolution of any primates, including, of course *Homo sapiens*, whose descendants are famous for similarly complex urban centers (McHarg 2006).

The probability that any planet has the temperature, pressure, and aqueous conditions that permit the biochemistry of life has been called a "life window factor" (Krumbein 2008). Julius Robert von Mayer, one of the founders of the science of chemical thermodynamics, wrote the following in his 1845 book *Die organische bewegung in ihrem zusammenhange mit dem stoffwechsel*:

Nature posed itself the task to catch light waving from Sun to Earth in its flight and to store this most motile of all forces in a solid form. To achieve this, nature covered Earth with the organisms of life that incorporate the sunlight and use this force to produce a continuous sum of chemical differences. (translation by W. E. Krumbein)

We suggest that interspecific sensory and coordinated physiological activity ("consciousness," as explained in this book, and "design," as explained in Turner 2007) are required for the construction of adequate and persistent "homes." One example is the coordination of sensory input, photo-autotrophic metabolism, and chemo-autotrophic metabolism in microbial communities that is required for transformation from two-dimensional flat laminae at environmental gradient surfaces (water-air, gravitational, organic chemical, and so forth) to form mineralized three-dimensional laminated biofilms (such as stromatolites). Further development into spheres and towers that induce oolites, pisolites, oncolites, and 6-meter-tall fluted Namibian termite mounds or Amazonian tree-top ant dwellings (Wilson 1991) requires far more detailed analysis. However, it is likely that "mindful" coordination of behavior, growth, metabolic, and genetic activities underlies such successful three-dimensional architectures as the magnetic termite mounds found in Australia, which are flat-sided like the tail plane on an aircraft and are aligned on a north-south axis to minimize the midday heat. Houses, nests, sheds, barns, food-storage structures, and other typical products of conscious organization involve coordinated activity by both conspecific populations and members of communities of species that, genetically, are related only remotely. Making a barn, to take an obvious example, requires not only humans but also trees. Here we limit our discussion to specific temperate-zone aquatic communities of oolitic limestone producers and metalworkers that shape certain underwater sedimentary deposits of iron and manganese.

As was summarized in the first paragraph of this chapter, ooids coordinate different genera and species of organisms into a logical, mindful, oolitic dwelling (Brehm et al. 2006; Krumbein 2008; Krumbein et al. 2003). The players are *Phormidium* sp., a cyanobacterium; *Navicula perminuta*, a diatom; and some less conspicuous chemo-organotroph bacteria, eleven in all, of which five are well-defined named taxa. These thirteen individual participants cooperate using all senses: chemoautotrophy, mechano-sensitivity, and photo-autotrophy (Brehm et al. 2006). First they need to reach a specific minimal number of individuals within the combined total population. Whether it is the larger associates (diatoms) or the smaller but more numerous bacteria that direct production and lithification remains unknown; most likely all are involved. In laboratory experiments, a chemical substance is generated within the laminated biofilm. The substance differentiates between higher and lower quantities of energy, carbon, and other sources. A complex film begins to be produced. Shaped as a sphere, a new layer or other repeat pattern not unlike the initial cells of a butterfly wing starts to grow. Then these nine to fourteen different types of organisms react to the substance, which in this context behaves as a signal or pheromone message. The cyanobacterial filament glides toward the spherically organized miniballoon and, if successful, penetrates the pellicle. Then the diatoms and heterotrophic bacteria accompany gliding cyanobacteria and their photosynthetic exudate. Together they penetrate the spherical barrier created by the community (figure 6.2). Inside the sphere a community pattern is observed by light and electron microscopy. Within the yellowish chemical globe several types of heterotrophic bacteria arrange themselves into a palisade. A tightly coiled layer toward the center contains the filamentous cyanobacterium *Phormidium hendersonii*. (See plate II.) CO_2-removing phototrophic and oxygenic cyanobacteria concentrate in the center of the translucent sphere, where they reproduce rapidly by cell division. They form multilayered rough or smooth spheres, some of which eventually solidify and calcify to form an organic-coated ooid. (See plate II.) A two-dimensional layered community transforms in this special way and becomes a spherical one, which creates its own

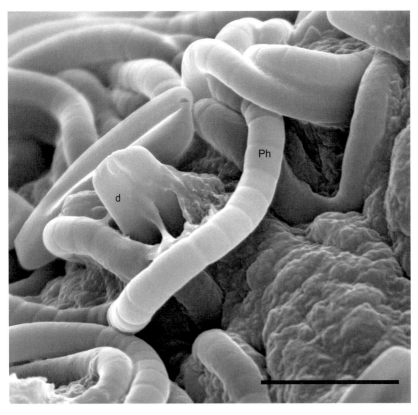

Figure 6.2
The surface of the sphere in figure 6.1 where the trichomes of *Phormidium* sp. (Ph) emerge; some move in and out. One diatom, d, a silicious alga, is "transported" or guided inside. Video documentation has shown unequivocally that only the cyanobacterium *Phormidium* sp. is capable of penetrating the spherical balloon "membrane." Motile heterotrophic bacteria and diatoms passively accompany the filamentous cyanobacteria inside, where the diatoms grow and reproduce. Eventually, spherically laminated carbonate replaces the silica of the diatom tests ("shells"). Scanning electron microscope image, scale bar = 10 μm.

chemical and metabolic gradients in a new dimension. If such metabolic growth, interaction, and reproductive activity developed into a wooden or a stone house, we would ascribe home construction to conscious behavior.

Not all organisms have such mindful organization. In our second example, ferromanganese nodules, the structures are a sign that organisms inhabit the environment. These microbialites are traces of microbial activity and, like ooids, are a product of the chemical gradient in the environment. Although the microbes that facilitate growth of nodules in the Second Connecticut Lake are not yet named and classified, we can extrapolate from analyses of comparable freshwater nodules in Oneida Lake in New York (Beliaev et al. 2002). Nodule growth is associated with the presence of *Shewanella oneidensis*, a Gram-negative heterotrophic bacterium that generates and removes intracellular iron and manganese oxides (Myers and Nealson 1988). Cyanobacteria, also present because of their phototrophic obligate oxygenesis, produce extracellular oxides of iron and manganese. The oxides form as by-products of free metal ions in solution. The nodule community nucleates the laminated sediments on solid substrates (e.g., rocks, sand grains, mud balls, even glass bottles).

In the Second Connecticut Lake (in New Hampshire, at the headwaters of the Connecticut River drainage system, which travels 655 kilometers to Long Island Sound), nodules are limited to an area of 100 square meters in the central western portion of the lake bottom at depths between 3 and 5 meters (Asikainen and Werle 2007). The hard but friable, generally laminated sediments display a wide variety of nodule morphologies in the small area. Most conspicuous and abundant are convex plate-like structures found in both "up" and "down" positions (figure 6.3A). At least three other nodule morphotypes supplement the plate-like nodules that form concentric rings around a nucleus: pavements, cup-shaped (<2 cm), and lattices. Underwater pavements derive their name from the continuous "sidewalk" formed by the growth of a porous pustular layer (1–2 cm thick) on the surface of plate-like nodules. In at least two localities, nodules join together into a single cohesive unit—a pavement (figure 6.3B). Small cup-shaped forms have the concentric ring pattern of the plates but lack an obvious central nucleus (figure 6.3C). Lattice structures, also without a central nucleus or concentric rings, have conspicuous voids throughout. They contain randomly distributed pebbles that range in size from small to coarse (figure 6.3D).

How rapidly do the nodules grow? Growth rate for nodules in the Second Connecticut Lake is estimated, from a 20-millimeter section of

Figure 6.3
Nodule morphotypes are intermixed within the nodule field. (A) Plate-type nodules with concentric ring patterns (scale bar = 2 cm). (B) Pavement-type nodules showing pustular morphology that coats individual plate-type nodules into one cohesive unit. (C) Lattice-type nodules. (D) Small cup-shaped nodules.

concentric-rings nodule material found growing on a commercial glass soft drink bottle discarded after 1930, to be about 26 millimeters per 100 years. If a constant growth rate is assumed, all the lake's deposits could have developed within 1,000 years. Growth rates of freshwater and oceanic nodules, calculated using radium (Ra^{226}) and beryllium (^{10}Be) dating, range from micrometers per year to micrometers per million years (Moore et al. 1980). If nodules take millions of years to grow to the size of a golf ball, we face a conundrum: Why weren't these structures incorporated into the sediment of the ocean or the lake? We contend that the assumption of unidirectional nodule growth is antiquated. We know that these metal-oxidizing bacteria both oxidize and reduce iron and manganese, depending on availability of both oxygen and metals. Thus, both buildup and breakdown of the metallic structures occur. It may not be practicable to measure growth rates accurately, since the reactions of consumption and reduction proceed in both directions to extents that cannot be measured with certainty.

Does ferromanganese nodule formation fit the idea of mindful organization? The details of modular "home structure" that corals (animals) or foraminifera (protoctists) construct are more intricate, yet the microbial community influences the size, shape, and durability of sediments within limited volumes in ways that are unprecedented in the abiotic geologic record (Schopf 2006).

Might spherical ooids show, by contrast with laminated nodules, the first signs of mindful community interaction? Who knows? In "home-building," including great height, humidification, temperature and solar radiation modulation, mineral and organic foodstuff storage, and defense systems, human achievements show continuity with microbial and other animal ingenuity over eons.

Bacteria began the same building trends: houses, gardens with fungal or plant crops in rows, tending of hedges and agricultural plots. We infer that denizens of the microcosm in community, bacteria, succeeded and supplemented by protoctists and animals, have built homes of some kind, supported themselves by "farming," and decorated and protected their turf with architectural structures of their own making ever since the anoxic Archean eon (Turner 2000).

7

Others

Laurie Lassiter

Whereas dogma once insisted that natural selection operated only "on the level of the individual," some current analyses focus on genes, populations, and species as units of selection. Here Lassiter introduces us to the work of Murray Bowen—a psychiatrist who never treated individuals in isolation, but always as members of human groups, specifically families. Bowen's theory of the family as a unit of natural selection involves a measure he calls "differentiation of self." The extent to which each of us responds to new social situations relatively independent of our early family experience is measurable.

The notion that individuals (or their genes) are the only bona fide objects of natural selection is deeply entrenched in neo-Darwinism. It is obvious, however, that human beings live, prosper, reproduce, and die in group settings—families, tribes, nations—that crucially affect the chances of an individual's survival. Moreover, the importance of the group to which the individual belongs, and in which he or she survives, is pan-biological: group membership matters not only to humans, but also to other animals and even microbes. Attempts to understand ourselves as individuals are doomed if we fail to take account of the overwhelming influence of the social groups in which we developed.

In most populations, individual organisms that live in the same place at the same time tend to form identifiable groups. The relationships between members of these groups often determine the health and fate of the entire group, even over more than a single generation. Murray Bowen (1978), a scientific investigator of human behavior for many years, recognized the transition from herds, flocks, and families of nonhuman animals to the tribes, clans, and families of humans. Years of studying the behavior of socially impaired, often institutionalized young people in the context of their families, along with later observations of "normal" families, led Bowen to document unnoticed regular processes of development in families based on systematic variation in siblings. He agreed with Charles Darwin (1898) that the "emotional system" (in this case the human family), not any individual person, is a minimal unit of natural selection in the evolution of the genus *Homo*. Bowen was especially interested in quantitative descriptions of automatic behaviors that involved four fundamental responses to social relationships. He noted that all people, to differing extents, use these responses:

distance, or avoidance of emotional pressure
capitulation, yielding, or giving in to emotional pressure from others
conflict, or refusal to give in to emotional pressure while attempting to force the other to capitulate
involvement of another person (often a vulnerable child) or persons through emotional pressure.

Bowen (1978) invented and refined a measurement called Differentiation-of-Self (DoS). Bowen's concept refers to the extent to which an adult responds to the environment based on perception and judgment (high score) or reacts as a member of his or her original familial group based on emotional programming (low score). The four types of responses, often hidden behaviors, can be observed after appropriate training. Behaviors of individuals may be predicted by means of Bowen family systems theory (ibid.). Not only do the analyses of Bowen and his colleagues apply to our species and to other social animals (e.g., chimpanzees, dolphins, wild and domesticated dogs); they seem to apply to microbes, including myxobacteria and heterocystous (nitrogen-fixing) cyanobacteria. Natural social phenomena such as altruism, aggregation into complex societies with shifting alliances, gang rape, mass deception, and crowd madness long antedate the appearance of primates, let alone humans. Some of this social behavior can even be detected in the fossil record. The dismissal of the observable phenomenon of "group selection" as not scientific unnecessarily turns a blind eye to a large and growing literature on the differential survival of animals as members of social units (families, tribes, or herds) relative to their more solitary close relatives. The worldwide expansions of both ants and humans are attributable to the effectiveness of their groups.

If the species *Homo sapiens* is unique, different from the rest of life, the study of human social groups—family, work, social organizations— is not related to behavioral science of other species. But if, as Darwin insisted, we humans evolved from nonhuman primates and, like them, are connected through evolution to common microbial ancestors, then human social behavior may be illuminated by study of other social species. Indeed I believe that scientific analysis of our social behavior contributes, in principle, to social studies of other animals and even of bacteria. Our behaviors have measurable antecedents in other forms of life. All life on Earth shares common ancestry, represented most closely today by the bacteria, as inferred from molecular biological and fossil evidence.

Murray Bowen was led to develop his ideas by psychiatric observations of repeated, predictable patterns of emotional behavior in the human family. In the middle of the twentieth century, he derived a widely applicable theory of human behavior from empirical studies. Bowen's inference of how individual behavior is regulated by social groups, especially the family, extrapolates to many other life forms. If predictable *Homo sapiens* social-development behavior were accurately documented in other species, including non-animal societies, a broadened social science would become seamless, in principle, as it integrated into the rest of evolutionary science. I argue that, even if incorrect, Bowen's theory of group behavior will generate observations that otherwise would not be made.

Current systems theories tend to be cybernetic analyses, and—whether applied to bacteria, nonhuman mammals, or humans—they mainly derive from engineering. Bowen, by contrast, directly observed human interactions that he theorized to be products of evolution. Believing that living systems could be accurately understood only by direct observation, he limited his systems theory to observable facts of living groups. In my view, his publications receive far less attention from scientists than they deserve, probably because they are confined to the psychiatric literature.

Murray Bowen, MD

Murray Bowen (1913–1990) grew up with an unusual ability to solve puzzles. As a youngster he accompanied his father—the only ambulance owner in Waverly, Tennessee—on a drive to Memphis as he brought a comatose girl to the university hospital. The confusion of the physicians led to her death within hours. Young Murray Bowen, who believed the girl's death had been avoidable, determined to become a physician. Interested in cardiac surgery, he built an early mechanical heart. On the basis of experience during World War II (which interrupted his medical studies), he concluded that doctors routinely found psychiatric casualties the most difficult to treat. Psychological treatment seemed the least effective of medical practice.

In Topeka, Kansas, Bowen joined Karl Menninger and others in an effort to bring a new science of human behavior—Freudian theory—to the United States. Eventually, however, he concluded that much of Freud's "theory" was unscientific and inadequate. Freud emphasized such undocumentable sources of emotional behavior as Greek dramas and patient narratives. Although Bowen learned from the theory and from its practice, studying it assiduously in a personal effort to move it toward science, he abandoned it by the end of the 1940s. He sought alternative approaches based on evolution and animal behavior.

Bowen maintained that hypotheses of living behavior must derive, not from engineering analogies, but from physiology and direct observation. He devised and executed long-term experiments—perhaps the only human field studies—at the National Institutes of Mental Health in Bethesda, Maryland. From 1954 to 1959, he and his team of nurses, psychologists, social workers, and psychiatrists observed entire families that had agreed to live on psychiatric wards for up to a year or even longer.

Bowen discovered that the family functions as an integrated whole, and that behaviors of which an individual is largely unaware are determined by the family unit. Bowen was more attentive to actions than to talk. Discrepancies between stated intentions and behaviors were routinely documented. What, Bowen asked, is the function of an action? He repeatedly noted that the effect of a behavior was the opposite of the stated intention. In one common example, a declaration of an intention to help a member of a person's family was followed by behavior that led to impairment of the family member's functioning. For example, a mother who insisted that her adult son should live his own life acted in ways that impeded, even precluded, his independence. These actions were seen as evidence of a systematic process that led to impairment of the function of some members of a family while enhancing the function of others.

Plate III includes a photograph of Dr. Bowen.

Bowen Theory

Bowen theory comprises several interrelated concepts understandable as aspects of a single phenomenon. One of these concepts is that individuals, largely unaware, are subject to and participate in systematic processes that occur in families or equivalent social groups. Most central is the recognition of the family as an "emotional system," a reproductive unit in which the behavior of any individual is part of a larger process. For Bowen, who saw human families in an evolutionary context, "emotional" means instinctual (in the broad sense). "Instinctual" or "emotional" here has the same meaning it had for Darwin when he referred to the emotions of man and animals. Darwin (1898) asked "whether the same general principles [of emotion] can be applied with satisfactory results, both to man and the lower animals." Since "emotional system" refers to repeated, systematic, coordinated behavior among regularly interacting members of a single population, I chose to retain the use of "emotional system" to name the same phenomenon in social groups across species, even though for bacterial colony populations it may seem odd. For Bowen, "feelings" were emotions and instincts of which people were aware and could speak. Systematic patterns of repeated behaviors of which people did not speak, because they weren't aware of them, were what most interested Bowen.

The degree to which any member is free from automatic response to signals in the emotional system is described, in Bowen theory, as Differentiation-of-Self. The lower the DoS, the more predictably a family member behaves, guided by stimuli of the emotional system. Especially during periods of perceived threat, coordinated behavior may contradict direct perception of reality as individuals attend to emotional signals from others in the group more than to facts in the social and the wider environment.

At lower DoS levels, individuals are conspicuously sensitive, often hypersensitive, to the behavior of members of their emotional system. They are preoccupied to a greater extent with how others feel about them, yet less aware of the extent of their emotional reactivity. The individuals most reactive to emotional cues from others are often those least aware of that sensitivity. Symbols, phrases, facial expressions, vocal tones, or other communication lead to automatic, predictable responses to emotional pressure. The survival and the reproductive success of any individual are subjugated to that individual's larger emotional system to some degree. But a trend is measurable: the lower the DoS level, the greater the subjugation of the individual to the family unit, and the more the individual will tend to participate in automatic processes that lead to his own and/or others' exploitation by the group. The exploitation of some by the emotional system increases the likelihood that the system as a whole will survive and reproduce, though certain members may not.

A mother who incessantly tends a mentally impaired or precociously talented child limits the reproductive future of the affected youngster but increases the freedom of his or her siblings to reproduce. Bowen described the processes that lead to the impairment of some members as reciprocal; that is, the impaired individual also participates, unaware, in the process that leads to his own impairment.

Once one has grasped Bowen's DoS concept, one can easily estimate variations in DoS levels by observing oneself and others. Bowen's DoS scale ranges from 1 to 100. Bowen believed that levels in principle were quantifiable and (especially with access to at least three generations' worth of family data) potentially scientifically objective. The extent of numerical-value consensus between Bowen and his colleagues at the Georgetown University Family Center (now the Bowen Center for the Study of the Family, where the Family Database Project defines the parameters of Bowen's scale) for an individual patient was remarkable when all members of the staff had independent access to the same set of data. Though their work was not validated in a formal research project, Bowen and his co-investigators often agreed within one numerical value—say, on 30 or 31—on the scale of 100. Bowen asserted that quantitative estimates of behavior—that is, of characteristics commonly believed to be personality traits—were valid.

Bowen observed four classes of behavior—distance, capitulation, conflict, and involvement of a third person (often a child)—used by everyone to manage lack of DoS in relationships that result in what he termed the "impingement" of one or more by the system. (See box 7.1.) He noted that those people who were assigned lower numerical values on the Bowen DoS scale were observed to utilize these behaviors with greater frequency and intensity, but there was no qualitative difference in behavior at different values.

The fourth class of behavior is due to the triangle's central function: to regulate the behavior of individuals in the emotional system. The triangle produces variation in DoS values—the child who is more emotionally involved in the relations of his parents, guardians, or siblings predictably develops a lower DoS value. Differential involvement of offspring in a family is the basis for variation in the development of DoS over generations. Intense emotional involvement leads to impairment in one or more person's social functioning. This often frees the other siblings to work and to reproduce.

The intensity and frequency of the four classes of behavior are determined by DoS values and by degree of reactivity to perceived threat.

Box 7.1
Management of lack of Differentiation-of-Self in relationships: Bowen's four classes of behavior of shifting alliances.

Distance, silence Imposition of distance (i.e., geographical separation) is used to avoid or reduce the frequency and intensity of emotional pressure. The lower the DoS, the more sensitive we are to emotional pressure and may seek to avoid it. At the same time, moving away we tend to re-establish similar social relations in new locations, as we also rely on emotional pressure to function. Distance is often comfortable for one person in the relationship and uncomfortable for the other. We may maintain physical closeness, live together in the same home, or even sleep in the same bed but nevertheless use distance by avoiding emotionally significant interaction.

Giving in, capitulation The individual who routinely capitulates to another is more likely to develop physical illness, depression, or other symptoms. A common fallacy is that of the "peace-at-any-price" person who believes he yields on important issues with no harm to himself. One who is catered to as others tiptoe around him, or one who gives up friends, interests, job, or goals and is cared for by others and becomes stronger as the one who sacrifices more becomes weaker, or one who is more isolated and is perceived as weak, vulnerable, crazy, or ill, or anyone repeatedly in a one-down position is a candidate for being taken advantage of automatically without his or her own awareness. Although Bowen and his colleagues reliably detected which of two spouses yielded more to pressures of their emotional system, that detection required training. Since the emotional system operates outside of awareness, often through unconscious lies and self-deception, first impressions and superficial analyses often are incorrect. The one who capitulates may be the invalid spouse who works less and less—whose health and functioning decline while the other takes more and more responsibility as his or her overall health and functioning increase. Alternatively, the one who yields may be the partner pressured to do more than his or her share as the relationships shift over time and circumstance.

Conflict, contentiousness, bickering Conflict occurs when neither individual is willing to yield to the other but each tries to force the other to give in. A common misperception is that conflicts occur because "he" is trying to force "me," while "I" am merely defending "myself." Conflict may range in intensity from occasional bickering to violence, depending on the DoS level and the degree of external stress of perceived threat.

Involvement of a third person, the triangle Parents or guardians may join together to discuss (and sometimes diagnose) a child and simultaneously relieve tensions in their adult relationship. A mother may complain about a child to the father. The father expresses anger to the child. The mother then defends the child, shifting the child's alliance to her and simultaneously keeping the father and the child at odds. A seasoned employee who complains to a new employee about the supervisor stimulates a negative relationship between the new employee and the supervisor that ameliorates the relationship between the supervisor and the seasoned employee.

When humans, like members of other social species, are exposed to threat, their groups react with increased automatic coordination as single units. Families, guilds, tribes, nations, and other social units tend to consolidate and to act as one.

The four classes of behavior describe emotional pressure on the vulnerable (the lower the DoS, the more observable impairment in functioning). Most families use a mix of the four, with impingement of several members predictably the result. One class of behavior may predominate in some families, leading to greater impairment in one family member. In families in which one of the spouses capitulates, for instance, and there is a lack of distance and a lack of conflict, with the children relatively free of the emotional process, the spouse who systematically yields to emotional pressure is likely to have a greater degree of impairment than in a family in which conflict and distance are more utilized, or in which one or more of the children are more impaired by the process (given that both families are at the same level of DoS). The higher the level of DoS, the less distance, capitulation, conflict, and involvement of a child occur. At the same level, families may utilize one or more of the four ways differently, but the behaviors will add up to about the same overall utilization of them.

Bowen recognized that these social interactions occur mostly outside of the awareness of the participants. His theory generates detection of processes that occur in people and their families—independent of nationality, social class, race, culture, or family size. In my work, I have observed the same processes in various family structures and across cultures. My clients have included people from Korea, India, Sudan, Egypt, Saudi Arabia, Venezuela, Lebanon, Italy, Pakistan, Spain, Mexico, and New Zealand, as well as from diverse backgrounds within the United States.

The Differentiation-of-Self Scale

Bowen showed a predictable correlation between numerical DoS value and potentially quantifiable frequency, duration, and intensity of the four kinds of social behavior. No correlation between DoS values and particular qualitative behaviors was found. In other words, people at higher DoS levels make use of the same four classes of behavior as those at lower levels, but to a lesser extent. All people manage their social relationships in those four modes.

No individual has achieved the maximum DoS-scale score of 100; it remains hypothetical. Those scored in the higher ranges tend to be guided more by individual perception and judgment. They assess social reality more accurately. Persons assigned higher DoS scores are more effective in maintenance of long-term, close relationships with others, and function more responsibly and flexibly. They tend to enjoy good health, productive work lives, and financial well-being, all things being equal. People who live in war zones or experience natural disasters often suffer difficulties beyond those indicated by DoS scores. Higher scorers often have vital relationships in close-knit families that can be documented for as many as four generations. They handle adversity with appropriate resolve, and they neither ignore realistic threats nor overreact to them. Freer from emotional reactivity to others, they display more freedom from the burden of the four classes of frequent behavior easily observable in those at lower DoS levels.

On a practical level, DoS refers to behavior not negotiable in social relationships. How extensively might DoS be interpreted? How free of reactivity to any environment might an individual be? How far might one go to base his responses on integrity rather than status in the relationship system? Interested in these questions, Bowen, who was not religious in the traditional sense and had not been raised in a religious home, referred to the poem attributed to Saint Francis of Assisi (box 7.2) as illustrating the substantive aspects of the DoS phenomenon. For those who practice Bowen theory, the poem can be viewed simply as a realistic, practical description of the effort to be free of reactivity to social pressures, and "dying" can be taken as neither a prelude to a supernatural

Box 7.2
A historical example of a high-score DoS: *The Little Flowers of Saint Francis* (adapted from an anonymous fourteenth-century Italian text).

Lord, make me an instrument of your peace;
   where there is hatred, let me sow love;
   where there is injury, your pardon Lord;
   and where there's doubt, true faith in you;
O Master,
   grant that I may not so much seek to be consoled as to console;
   to be understood, as to understand;
   to be loved, as to love;
   for it is in giving that we receive,
   it is in pardoning that we are pardoned,
   and it is in dying that we are born to Eternal Life

afterlife nor an admonition for ideal behavior, but simply changing oneself to adjust to reality.

Bowen, who considered the highest DoS scores to be probably hypothetical and not achieved, asked himself if any historical figures might have exemplified such high DoS values or might have achieved a significant increase in DoS over a lifetime of living by principle. Saint Francis, who was emotionally close to his mother but estranged in childhood from his father, may be an example of one who raised his DoS value when evaluated by his behavior over a lifetime. Raised in what is now Umbria, Francis (ca. 1181–1226) abandoned the opportunities of his social class to become a friar. He established the Franciscan Order and was known for his ability to gain acceptance even from those who initially opposed him. He dictated the following in 1220:

Francis said: "Friar Leo, write." Who responded: "Behold I am ready." "Write"—he said—"what is true joy? A messenger comes and says that all the masters of Paris have entered the Order, write, 'not true joy.' Likewise that all the prelates beyond the Alps, archbishops and bishops; likewise that the King of France and the King of England: write, 'not true joy.' Likewise, that my friars went among the infidels and converted them all to the Faith; likewise that I have from God this grace, that I heal the infirm and work many miracles: I say to you that in all these things there is not true joy. But what is true joy? I return from Perugia and in the dead of night I come here and it is wintertime, muddy and what is more, so frigid, that icicles have congealed at the edge of my tunic and they always pierce my shins, and blood comes forth from such wounds. And entirely with mud and in the cold and ice, I come to the gate, and after I knock for a long time and call, there comes a friar and he asks: 'Who is it?' I respond: 'Friar Francis.' And he says: 'Go away; it is not a decent hour for traveling; you shall not enter.' And again he would respond, insisting: 'Go away; you are a simpleton and an idiot; you do not measure up to us; we are so many and such men, that we are not in need of you!' And I stand again at the gate and I say: 'For the love of God take me in this night.' And he would respond: 'I will not! Go away to the place of Crosiers and ask there.' I say to you that if I will have had patience and will not have been disturbed, that in this is true joy and true virtue and soundness of soul.

In the absence of dogma, religious injunction, and moral judgment, Bowen, on the basis of his keen observations and scientific approach, apparently concurred with Francis's recognition of the value of emotional freedom in the face of adversity.

DoS levels are measured as distinct from other factors that may affect social relationships and overall functioning. They do not correlate with intelligence, talent, innate personality traits (such as shyness or extroversion), cultural style, socioeconomic class, or political or economic circumstances. They are largely independent of genetic propensities. The major determinant of an individual's DoS score is the history of the individual within the family and the functioning of the emotional system in which his or her social development occurred. To ascertain a numerical score would require, at a minimum, knowledge of the individual's social history (i.e., relationships with parents, siblings, spouse(s), and offspring) and of the documented functioning level of each, including at least three generations. Bowen theory posits that the basic nature of an individual's historical relationships with others, especially early experience with parental figures, determines the DoS, that the DoS is the major indicator of an individual's physico-mental health, and that it pertains not only to family relationships but also to social relationships in work life and in other areas.

People with lower DoS values show increased sensitivity to emotional pressure, more frequent, more intense, and longer-lasting emotional reactivity, and less emotional flexibility. They tend to be upset by minor disturbances. They are more self-absorbed or more intensely focused on one other person. They expend less energy on behavior beyond variations of the four classes of management of lack of DoS. Some use intellectual activity, often with a high emotional charge, to help manage their social relationship difficulties. Habitual behaviors that aid in emotional management tend to extremes: self-sacrifice or selfishness. Those with lower DoS scores tend toward increased reactivity. They respond emotionally, with anger, hysteria, stubborn silence, or other inappropriate habitual behavior, to relieve their personal stress in the present moment. They fail to act on principle or to plan ahead. Often, to avoid emotional pressure, they flee their primary emotional system, family of origin, and/or nuclear family. Since they cannot function without the intense pressure of their emotional system, however, they predictably replicate it in new relationships. Those with the lowest DoS scores require continual guidance by external pressures of the emotional system. Having failed to develop a guidance system within themselves, owing largely to factors within the emotional system in which their development has occurred, many of the lowest-DoS people are so reactive and so unable to cope responsibly that they must be institutionalized in a prison or a psychiatric hospital.

Bowen distinguished what he called "solid self" from functional level, or "functional self." DoS is independent of immediate circumstance— and independent of negotiation within the system. "Functional level," especially for those with lower DoS levels, varies over time and is contingent on situations, such as making a new friend, receiving an award or a promotion at work, losing a job, being divorced, or being rejected by a lover or a boss.

Functional level may shift through reciprocal relationship processes. People "borrow and trade" functional levels through stereotypical predictable behaviors of which they are not aware (Bowen 1978). Among members of an emotional system, some may enhance one's individual function by preoccupation with another's perceived problem. One enlists others to focus on the problem as well. Borrowing and trading decreases one's functional level, with simultaneous reciprocal increases in others. The four classes of behavior (distance, capitulation, conflict, involving a third) regulate social relationships to stabilize the emotional system in a way that results in some individuals', as a result of being impinged upon, functioning more poorly. The inevitable borrowing and trading in relationships is exploitive of some, while benefiting others. Bowen thought that if one could become aware of his or her propensity to participate habitually in the behaviors listed in box 7.3, the self-awareness itself might help thwart unpremeditated harmful action.

The Triangle Hypothesis

Individual behavior is guided by the emotional social system. The individual's intimate group, usually the family, guides responses to many stimuli in ways of which the individual usually isn't aware (Bowen 1978). The "triangle hypothesis" (an abbreviated way of saying "the regulatory-function-of-the-triangle hypothesis") proposes systematic, identifiable behaviors that occur in social animals, and even in some version in microorganisms, including those that most closely represent early life—bacteria. I developed the triangle hypothesis over 15 years of applying the Bowen theory to biologists' observations of social species and studying the direction of Bowen's research in his last decade, using notes and audio tapes of his meetings with individual patients.

The triangle hypothesis offers an explanation of how individual members of a social species force other members to contribute to the survival and reproduction of the group in ways that may even threaten themselves. In contrast to the unfounded neo-Darwinian assumption that only individuals are units of natural selection, the triangle hypothesis recognizes that the family or some other group (that is, the emotional system) is a unit of natural selection. In fostering the overall survival and reproductive success of group members, the emotional system exerts pressure on all its members, some more than others. During times of

Box 7.3
Behaviors more frequent and intense in individuals with lower DoS scores.*

1. Blame others for problems, but lack awareness of own responsibility
2. Avoid discussion of emotionally significant topics until upset
3. Easily pressured into acts of selfishness or self-sacrifice; and pressure others
4. Take sides on emotional issues, lack neutrality
5. Respond to remarks or observations from others as if they were personal criticism
6. Quick to blame others or themselves even in the absence of evidence
7. Worry excessively about the impression they make on others
8. Criticize, exaggerate the faults or abilities of self or others
9. Avoid social gatherings and emotionally challenging interactions
10. Idealize themselves or others
11. Distort actions of others, making exaggerated assumptions based on personal wishes or fears
12. Assume unwarranted responsibility for the happiness, well-being, or ruin of others

*This is a numbered but not ordered list of overlapping behaviors that fall within the four classes (distance, conflict, capitulation, involvement of a third person—see box 7.1). They are not to be confused with inherent personality traits. Rather, these are social-group phenomena that develop anew in each generation within the emotional system in response to social and other environmental variables. A function of heightened social anxiety associated with these behaviors is the control of members inside a given system. To the degree that individuals are regulated by their immediate emotional system, they may be used by it—to be exploited themselves or to exploit others in close contact. These behaviors are best interpreted to be activities that ensure individual service for group survival and reproductive success.

scarcity or some other threat, in coordinated behavior each member is pressured by the family, and, simultaneously, each puts pressure on the other members. The triangle hypothesis is an idea about shifting alliances with explanatory power that generates new observations and suggests future experiments in social species.

After Bowen's research led him to identify the "triangle," which I have described as an automatic two-or-more-against-one behavior system, he used his knowledge of triangles to raise the DoS levels of his patients or those of members of their families. His technique increased individual regulation vs. regulation by the emotional system and generated new observations about the emotional system. Although his work did raise the DoS levels in patients and their families, I believe that Bowen wanted to understand the dynamics of the emotional system more than he wanted to practice therapy.

In the 1950s, while ostensibly researching schizophrenia at the National Institute of Mental Health, Bowen extended Darwin's idea to show how the emotional system of the human family, like that of other groups of social species, is a product of evolution. He kept that work private to avoid the interrupting controversy that he reckoned would be inevitable. And during the last ten years or so of his life, Bowen kept his clinical research private to protect his work from the vociferous criticism it generated among colleagues who valued only effective therapies and had no interest in or patience for Bowen's scientific pursuits. His early attempts to coach people more openly had led to some confusion and to some emotional reactions against his techniques. In the 1980s he let it be known that he would terminate his coaching of colleagues.

Notes and audiotapes of Bowen's experimental sessions document how "triangles" control individuals, who themselves also participate in alliances to control others, beyond their own awareness. When an individual takes steps to self-direct, hidden triangle threats activate predictable group behaviors to regain control of that individual. The emotional system, functioning for the survival of the group, has evolved to react so as to regain control of the individual who attempts self-guidance. The triangle threats take the form of taking sides against the individual, while invitations are made to join the emotional system. Exaggerated threats and invitations make visible what is usually hidden.

Shifting alliances occur, each member seeking a comfortable position in relation to others. Humans predictably seek one of the two or more triangle "inside positions," each wanting approval, acceptance, and reassurance from the others. Each opportunistically uses the two-against-one threat of rejection to exert pressure on the others. The emotional system ensures the family unit's survival and therefore increases the probability of reproduction by regulation of its members. All tend to be unaware of the regulatory function of their own responses.

The intensity of cues (both invitations and threats) inherent in the group behavior correlate with the degree of a member's inability to guide himself independently without reliance on frequent cues. The inability to initiate one's behavior, accompanied by predictable responses that seek chronic reassurance and chronic praise and avoid rejection by emotional unit members, generates anxiety (= stress reactivity). The lower one's DoS score, the greater is his or her sensitivity to the triangle. The greater one member's need to be in an "inside" position, accompanied by reactivity to expulsion from such a position, the greater is that member's chronic anxiety and stress reactivity. The triangle hypothesis predicts the frequency and the intensity of stress responses in social relationships.

If excessive reactivity impairs group members in social relationships, why does it persist? Excessive sensitivity, usually detrimental to an individual, functions to optimize the probability of survival and reproduction of the group.

All humans develop in complex social groups whose members vary in DoS. Higher DoS is desirable. The higher the number of individuals within a group who respond to the reality of the fluctuating environment, rather than to habitually responded-to emotional pressures, the more effective is the group. Yet DoS is produced through development in the family within a sibling group in which some siblings develop lower DoS, freeing others to develop higher DoS. Bowen theory explains the wide variability in functioning and health that can be observed in *Homo sapiens*—a range that can be noted in all families over multiple generations—and offers ways to increase DoS by awareness that minimizes tendencies to impinge upon group members.

Group Selection in Non-Human Animals

Does Bowen's description of the triangle apply to species other than *Homo sapiens*? I suspect, as did Bowen, that these human behaviors are so basic that they will be found, if sought, in non-human social animals and beyond. Bowen-theory concepts generate potentially new scientific research. Knowledge of social behavior of human and non-human life is complementary; advances in understanding cross-fertilize the study of both.

New facts about social mammals have been coming to light. Two-against-one processes that have regulatory functions have been identified in non-human species. Only since the 1980s, research of dolphin social behavior has discovered complex alliances that control mating and hunting (Connor et al. 2000). Chimpanzees form alliances that determine troupe dominance (De Waal 1982). Though these alliances that resemble two-against-one patterns in the human triangle are not necessarily ubiquitous and require more research, regulatory processes that are equivalent in function to the triangle may occur in all social species.

In the 2005 documentary film *March of the Penguins*, hundreds of male penguins unite in a circular mass on an Antarctic ice sheet in

December for two months of egg incubation. They do not feed while they keep the eggs warm. Male penguins appear to take turns in an orchestrated dance so that each has an opportunity to be both in the center of the mass and outside it. Although it is known that not all of the penguins survive and successfully reproduce, individual variation based on social relationships in the mass of male penguins has not been studied very much. Penguins in the central, warmest part of the mass are protected from the bitter cold. Do some get pushed more to the circumference—or push others outward in an attempt to get into the center? Loners would predictably perish, and if some relative loners found themselves more often toward the outside of the group then others would be permitted more time toward the inside. The cold induces severe stress, which threatens the survival of both eggs and male penguins. During periods of increased threat to the group, the emotional system increasingly guides individual behavior, and impingement of some may accompany survival and reproduction of the larger unit.

Cellular Slime Molds

In *Dictyostelium discoideum*—cellular slime mold—the individual amoeba reacts to the pheromone (social hormone) cyclic adenosine monophosphate (cAMP). During periods of stress—such as desiccation or starvation—cAMP production is induced, and all amoebae stream toward local concentration maxima. (This social hormone of bacteria is discussed in chapter 3 of the present volume.) Each amoeba excretes the hormone, and each reacts to it by moving toward the highest concentration that it perceives. The system regulates the individual as amoebae stream toward the highest concentration in a central aggregation. Independent of genetic relationship, individuals respond to concentration gradients produced by their fellows. Those with the greatest reactivity excrete the largest amount and race toward the position that becomes the dead stalk made up of individual organisms that now function like cells in the multicellular structure. Those with lower reactivity to the social signal excrete less cAMP and develop into the propagules. Depending on the stalk for dispersal into a new environment with needed nutrients, these spores enter the future generation. Amoebae in the pre-spore position of the structure move in a deliberate, orderly manner. In the pre-stalk position, movement is chaotic and hyperactive; with a greater number of surface pheromone receptors, these individuals show greater reactivity to the social cues and do not survive to the next generation (Bonner 1998).

The hyperactive amoebae resemble those humans who develop the greatest reactivity to the alternating approval and rejection in the two-or-more-against-one triangle. Yet "a three-position process" does not describe what in cellular slime molds determines that some amoebae become dead stalk and others become the living spores. The word "triangle" even misleads in describing human social behavior. Even the hypothesis of groups in which some individuals gang up against others in shifting, opportunistic alliances is inadequate when applied to all social species. The human, species-specific two-or-more-against-one triangle exemplifies a broad category of social group regulation. The hypothesis that all social species are regulated by their group members generates detailed observations.

Regulation of individuals by emotional systems is not yet systematically studied in biology, although descriptions of individual variation in social groups ranging from bacterial colonies to baboon troupes provide facts that support it. The neo-Darwinian tradition of evolutionary thought that rejects "group selection" out of hand lacks a theoretical approach to social living. Life science itself lacks a theoretical approach to integrate studies of automatic, predictable social (group) behaviors that become more intense under stress. Bowen adds something new to the study of life: the idea of an emotional system ("emotional" in the sense in which Darwin used it, meaning instinctual) that regulates individual organisms, along with variation in degree of group regulation of individuals (DoS).

Microbe Sociology

It is the study of group behavior in bacteria, of all social organisms, that has been most open to a systems view (Shapiro and Dworkin 1997). The small size of bacterial individuals and their groups—as well as, perhaps, their presumed irrelevance to human social systems—has left researchers free to evaluate how the group determines the life course of a single individual. Ideas about individual human behavior that bias the study of animal societies, especially other primates, is much less marked in these newer studies of the social systems of bacteria.

Cyanobacterial Altruism

The application of Bowen theory to the simplest and earliest life forms, from which all others evolved, reveals what is most basic about social relationships in general; simultaneously it offers new ways to think about bacteria. Filamentous heterocyst-forming cyanobacteria evolved more than 3,000 million years ago. Living and ancient multicellular bacteria show the same basic morphology. Extant cells observable today greatly resemble the fossils seen in thin sections of rock. The social and reproductive group is the beautiful blue-green cyanobacterial filament. About 10 percent of the individual cells that make up the filament transform under stress to live for the group. These specialists are large non-reproductive cells called heterocysts. (See panel B of plate III.) Unable to photosynthesize and provide themselves food to maintain and grow, heterocysts cannot reproduce. However, they remove inert nitrogen from the air and convert it to a chemical form that can be metabolized by members of their "emotional system" (the reproductive cells in their filament).

When *Anabaena* sp., a filamentous cyanobacterium, experiences a deprivation of ammonia or nitrate, its usual nitrogen source in the soil or water, the bacterial filament—the reproductive and social group of the cyanobacterium seen as a chain of cells—changes. The change leads some individual cells, at regular intervals along the filament, to transform: the heterocyst's transformed wall now blocks the entry of oxygen that would otherwise poison the nitrogen-fixing enzymes. The heterocyst's blue-green color is replaced by brown when the chlorophyll of photosynthesis is no longer synthesized. Filament cells that do not transform continue photosynthesis; they remain green, release oxygen, feed themselves, and reproduce by division. (See panel C of plate III.) They survive on the usable nitrogen produced in brown heterocysts.

That automatic programmed altruism exists in these cyanobacteria was shown dramatically by Popa et al. (2007). Organic (usable) nitrogen produced by the non-photosynthetic, non-reproductive mature heterocyst is altruistically transferred to its photosynthetic filament neighboring cells (incompetent to fix nitrogen) within 90 seconds of the metal-ion-mediated stimulus. (The metal that mediates the stimulus is manganese.) The stimulus to break the very strong $N \equiv N$ bonds and incorporate atmospheric gas into organic matter stimulates only the transformed heterocyst cell, which, within a few minutes, transfers most of its organic nitrogen to its green, photosynthetic, growing neighbors, keeping nearly none for itself. The heterocyst never grows or reproduces again. It suffers programmed cell death.

How does one cell sacrifice itself and become a heterocyst? A smaller cell, or any other in the filament more sensitive to nitrogen starvation, reacts to scarcity earlier than larger, less sensitive cells. Once a cell transforms into a heterocyst and leaks organic nitrogen to its neighbors, it relieves other nearby cells of the need to transform. All cells are genetically capable of conversion to heterocysts if nitrogen starvation pressure suffices. A cell that begins to transform into a heterocyst can revert back to a photosynthesizing and reproducing cell only if an adjoining or nearby cell at the very beginning of its transformation reduces local stress by relieves the pressure by completion of transformation into a heterocyst. Once the heterocyst develops, reversion is impossible.

Heterocyst development is regulated by precise external environmental conditions where one cell's environment includes the rest of its filament. The cyanobacterial filament, the social and reproductive unit, is capable of regulating the genes of other cells in the filament. The filament is the "emotional system," a functional unit, a group that survives to the next generation.

The early appearance of fossil heterocysts along filaments of once-green cyanobacterial cells raises questions. Is the formation of impaired, deficient, specialized members of a group induced by the group itself correlated with survival of the group? Might Bowen have discovered a general characteristic of social beings? Life requires growth and energy transformation and, to ensure continuity via reproduction, in most animal species individuals tend to survive only in functional groups, such as mating pairs, families, herds, tribes, villages, or colonies. All life is stressed by paucity of food, energy, or space and by other environmental threats. When individuals behave so as to enhance their group's survival, does the larger emotional system survive and reproduce more predictably than do related but isolated individuals? Was the emotional system Bowen describes in human families present in early life forms? Growth of cells leads to reproduction in all of life—might epigenetic behavior of the group be prerequisite to survival and reproduction in social species? Observation of such group behaviors leads me to reject the insistence of zoologists that no group selection exists.

What behavioral, metabolic, physiological, and genetic processes lead members to behave against their own survival and reproductive interests in ways that promote the survival and reproduction of their group? Certain individuals are more vulnerable than others to elimination by natural selection because of their smaller size, younger age, distal position in the group, or some other circumstance. These individuals are rendered more reactive to signals from their social surroundings. Bowen's main idea—that developmental processes in the family emotional system determine individuals' life course trends—helps explain extremes in survival and reproduction in social species. Bowen's more inclusive perspective is conducive to detailed analysis of individual and group behavior for accurate description of the social lives of not only the human family but also the many other gregarious forms of life.

8

Nested Communities

James MacAllister

As only one of an estimated 30 million extant species on a planet where more than 99 percent of the species that ever lived are extinct, we short-sighted humans are pathologically preoccupied with the present. Here MacAllister examines two scientific concepts that extend our view of ourselves: Gaia (the idea that Earth's surface forms a continuous physiological system) and symbiogenesis (an evolutionary process that may lead members of different species or higher taxa that live in physical contact to ultimately merge physiologically and genomically to generate new organisms and species).

Western, institutional science has given either no reception or a negative reception to crucial information showing that evolution results from individual, social, and communal interactions at scales ranging from the microscopic to the planetary. Here I briefly review two recently developed ideas. The first is Gaia, the second symbiogenesis in the context of community. Over time, community ecology becomes evolution, and organisms internalize their external environment (Margulis et al. 2000). I begin with powerful, anthropocentric cultural metaphors in which, unawares, you and I are steeped: "Spaceship Earth" and "selfish genes."

Gaia Yes, Spaceship Earth No

James E. Lovelock (b. 1919), the British atmospheric chemist who invented the scientific Gaia hypothesis, named his idea (with help from his neighbor, the author William Golding) for the ancient Greek Earth goddess to emphasize that Earth isn't just a pile of rocks to be kicked; rather she is a living planet to be respected. The scientific Gaia hypothesis postulates the following:

Over 30 million types of extant organisms, descendant from common ancestors and embedded in the biosphere, that directly and indirectly interact with one another and with the environment's chemical constituents, form a biotic-planetary regulatory system. They produce and remove gases, ions, metals, and organic compounds through their metabolism, growth and reproduction. These interactions in aqueous solution led to modulation of the Earth's surface temperature, acidity-alkalinity, and the chemically reactive gases of the atmosphere and hydrosphere. (Margulis 1998)

"Spaceship Earth" is a grotesque mixed metaphor. According to Lovelock (2006),

Metaphor is important because to deal with, understand, and even ameliorate the fix we are now in over global change requires us to know the true nature of the Earth and imagine it as the largest living thing in the solar system, not something inanimate like that disreputable contraption 'spaceship Earth'.

"Spaceship Earth" assumes that we humans are the planetary flight crew and the rest of life is just along for the ride. But we humans act more like the only first-class passengers. On gravitational autopilot, we circle the Sun at that "just right" distance. Our swelling human numbers force the remaining non-human passengers into the last few seats left in coach. We eat and drink like gluttons. Now, the galley refrigerators are defrosting and the frozen entrees have begun to spoil. We are oblivious to the plight of non-humans, not realizing that they are our life-support system. Constant human fights over who merits a window or an aisle seat, and who gets pretzels or macadamia nuts, have already trashed the passenger compartment. The lavatory's smoke detectors are disabled. We just ran out of bottled water and toilet paper. Our waste has overflowed the toilets; now a fetid slurry of blue chemical, urine, and excrement spreads across the floor and down the aisle. We have encountered turbulence, but no one heeds the illuminated "FASTEN SEATBELT" sign.

A Test for Martian Life: Lovelock's Gaia

Lovelock (a chemist, as I have already noted) was employed by NASA in the late 1960s to design an instrument for the Viking mission to seek Martian life. Lovelock objected to the assumption that Martian life would resemble the sugar-metabolizing or chlorophyll-bearing photosynthetic life of Earth. Aware that any life form capable of exponential growth cannot hide its energetic and material needs, he developed an elegant test. Life must use fluid media, liquids and/or gases, to transport the energy and matter it requires. A lander wasn't necessary; one had only to measure the gas composition of the Martian atmosphere from Earth. If, on the basis of the laws of chemistry and physics, the mixture were wildly improbable, the presence of life could be inferred. If atmospheric gases were entirely consistent with chemical principles (as on Mars and Venus today), one would infer the absence of life.

Planetary Anomalies

Struck by the contrast between the energetically spent CO_2 atmospheres of Mars and Venus and the reactive gas mixture continuously available to do work on Earth, Lovelock recognized that he had unwittingly detected life on Earth. On Earth, methane (CH_4) and nitrogen (N_2) persist in the presence of oxygen, with which they react. These gases are all products of certain life forms. Furthermore, the enormous discrepancy in the amount of water on Earth (3,000 meters if spread over the surface area of the planet evenly) relative to the scarcity on Mars and Venus (≤ 1 meter) impressed Lovelock (Harding and Margulis 2009). Earth's moderate surface temperature also required scientific explanation. Lovelock knew that astronomical models of stellar evolution showed that radiant energy increases as stars age. The Sun is 20–30 percent more luminous now than it was in the early Archean eon (4,000 million years ago). Earth's surface temperature remained between 0 and 100°C throughout the period of the fossil record and probably was even more closely constrained— between 5 and 30°C. Earth's temperature, Lovelock concluded (1988), is actively regulated. Again, indirectly, he had detected life on Earth.

Gaia is a self-starting system with cybernetic tendencies. It has limits within which life is constrained. It also has "set points" (specific temperature ranges, gas concentrations of chemically reactive mixtures) around which negative feedback loops work toward a dynamic equilibrium. One example of a Gaian negative feedback loop is production of dimethylsulfide (DMS) gas by ocean-dwelling algae (mostly a planktonic haptomonad *Phaeocystis*). Over the oceans, DMS released into the air reacts to form tiny droplets of sulfuric acid that may become "condensation nuclei" for ocean clouds. Clouds reflect more solar energy than the ocean surface and thereby cool the water below. The algae then make less DMS. Lower concentrations of DMS leads to fewer clouds; more solar energy is then absorbed, and local temperatures rise (Harding 2009).

Earth is an open thermodynamic system that runs on an incessant input of energy from the Sun. Energy flow across gradients generates organization and order. (See chapter 21.) Pressure, temperature, and radiant energy gradients produce tornados and hurricanes. These also appear to be driving forces that underlie and organize "life"—a process that is at least 3,000 million years old. More than 30 million named species, strains, and varieties—perhaps more than 100 million kinds of life, all sharing common ancestry—are estimated to be extant today (Schneider and Sagan 2005).

Inconsistent with Everything

When Lovelock proposed his Gaia hypothesis, scientists ignored or disparaged the idea.

Every scientific idea passes through three stages, as the English scholar William Whewell noted in 1840 in his masterpiece, *History of the Induction Sciences* (Kozo-Polyansky 1924 [2010], p. 2). First, it is ridiculed. Second, it is violently opposed or claimed to be of only minor importance. Third, it is accepted as self-evident.

In "Goddess of the Earth," a television documentary produced by John Groom for the BBC in 1986, the Oxford University zoology professor Richard Dawkins says of Gaia: "It's not exactly distressing or disturbing, except to an academic biologist who values the truth. I'm quite happy to say the function of a bird's wing is to keep the bird up, the function of the bird's eye is to form a well focused image. I'm quite happy with that kind of purposeful language because that's at the right level in the hierarchy of life. What I'm not happy about is to talk about the function of a particular gas in the regulation of the biosphere because it implies that individual organisms manufacturing that gas are doing it for the good of the biosphere."

In 1981, the molecular biologist Ford Doolittle of Dalhousie University wrote of Lovelock's Gaia hypothesis: "These ideas are inconsistent with everything we think we know about the evolutionary process. I do not doubt that some of the feedback loops which Lovelock claims exist do exist, but I do doubt that they were created by natural selection or that they are anything but accidental." (Doolittle 1981)

In 1983, Lovelock and his former PhD student Andrew Watson invented a simplified model of a world with only one species of plant: daisies of two varieties, white and black. Its star, similar to the Sun, grows hotter over time. Daisies grow between 4°C and 40°C, with an optimum growth rate at 20°C. As the temperature rises, black daisies absorb radiant energy, grow to larger population densities, and thus warm their local habitats. As the temperatures continue to rise, the smaller white daisy population, which reflects more sunlight, expands to cover a greater area and thus cools the local surface. The growth and reflection response to rising temperatures generates temperature control within the tolerable temperature range. Once the hot temperature is beyond the range of the daisies' response, they die, the temperature modulation disappears, and the planet's surface temperature, like that of the other planets, is directly determined by increase of solar luminosity and distance from the star. The "Daisyworld" model demonstrates that, in principle, Gaian regulation over geological time is achieved by response to a warming sun, temperature sensitivity, the response of differential exponential growth rates of varieties in a population to local temperatures, and the mix of reflective surfaces of black and white daisies. Conscious purpose is not involved. What Dawkins, Doolittle, and other evolutionary biologists fail to grasp is that among Gaia's emergent regulatory properties several features they assume to be intrinsic to evolution are not required—e.g., hierarchy, competition, struggle, evolutionary change generated by random mutation, even natural selection.

Lovelock's Gaian view originally posited that temperature, the atmosphere's reactive chemical composition, and the acidity or alkalinity of the ocean are modulated by life (Lovelock 2006). Extensions of the Gaian idea are under study. Do life forms redistribute elements (e.g., metals, such as manganese or gold) or rocks (e.g., limestone, pumice, granite, basalt, salt, ice) in ways that influence themselves and other life forms? Is the salinity of the oceans kept below saturation levels by life? In evaporite basins, does microbial life retard salt dissolution? Does life augment abiotic geological rates of interaction among solid, liquid, and gas interfaces? Bacteria eat tunnels into oceanic basalt, mostly glass, which augments surface area available to weathering. Does life act to retain water, relative to Mars and Venus, which have lost so much of theirs? Does life redistribute water on a planetary scale? Is the production of granite (0.4 percent of the lithosphere's volume) a Gaian phenomenon? If so, we would expect this widely distributed continental rock, composed of quartz, feldspar, hornblende, and mica, to be absent from all lifeless planets. So far, granite has not been found elsewhere in the solar system. Limestone, water, and granite are related to lithosphere upheaval, plate tectonics, volcanism, and mountain building. Lateral movement of tectonic plates may also be a Gaian phenomenon. (See chapter 10.)

Symbiogenesis

Symbiosis—a term invented by Anton de Bary in the nineteenth century to refer to physical contact between two or more "differently named" organisms that lasts for most of the life history of at least one of them (Margulis 1998)—never did imply social relations or mutual aid. Symbiotic relationships, behavioral and metabolic, over evolutionary periods of time, may generate morphological and metabolic change in either or both of the partners, and eventually even genetic dependencies. *Symbio* (living together) *genesis* (from *genos*, origin), an evolutionary concept, refers to such detectable changes in the associates as new behavioral traits, new tissues, new organs, new organelles, and even new species and more inclusive taxa (genera, families, classes, or phyla). Russian literature has documented "symbiogenesis" for more than 100 years. K. S. Mereschkhovsky (1869–1910) coined the term, and other Russian symbiogeneticists, A. S. Famintzin (1845–1905) and B. M. Kozo-Polyansky (1890–1957), described many examples of symbioses in nature (Khakhina 1992; Kozo-Polyansky 1924). Mereschkhovsky argued that joined forces—symbiotic partnerships—produce dramatic solutions for survival. Lichens—symbiotic associations of fungi (mycobionts) with photosynthesizers (photobionts), either algae or cyanobacteria—are estimated to cover 8 percent of Earth's land surface. The lichens produced by these mergers are unique new organisms with little resemblance to either associated partner. Lichens thrive on bare rock, on dry bark, in new lava fields, and in other habitats where neither the fungus nor the photosynthesizer survives on its own.

Darwinist Descent with Modification

Like all animal husbanders and gardeners before them, Jean Baptiste de Lamarck (1744–1829), Alfred Russel Wallace (1823–1913), and Charles Robert Darwin (1809–1882) were aware that offspring both resembled and differed from their parents. This observable difference is "inherited variation." Neither Darwin nor Wallace nor anyone else in the nineteenth century knew what caused heritable change. Darwin (1868) suggested pangenesis, the inheritance of acquired characteristics associated in the English-language literature with Lamarck. In Darwin's scheme, new physiological characteristics (e.g., larger muscles generated by exercise) produced gemmules (heritable particles) that collected in the "germ plasm" (sperm and egg cells).

All populations of organisms have a measurable "biotic potential": the number of offspring that under unrestricted conditions of matter and energy flow can be produced per generation. Maximum reproduction rates (offspring per generation) are species specific. They vary from one cell division in 15 minutes to produce two cells, as in certain bacteria, to human populations, in which one couple can produce more than 20 children in 30 years. The potential exists for all biota to produce more offspring than their local environment can support. Since not all off-spring live to reproduce, the result is Darwin's elimination process: natural selection. Those that survive tend to pass their traits to their offspring. Darwin and Wallace proposed the idea of descent with modi-fication Over long geologic time scales, this gradual, generation-by-generation process produces new and different varieties of organisms that diverge from their ancestors. But Darwin and Wallace, their prede-cessors, and most of their successors never answered in detail the ques-tion of how any new species originate.

Darwin insisted that descent with modification is gradual, yet the fossil record shows that well-documented plants and animals with excellent preservation potential appear suddenly and discontinuously. Organismal lineages are punctuated intermittently by speciation. Not only do members of species not appear gradually in a series of transitional forms; they disappear in sudden extinctions. *Gradual* diversification of species in the fossil record of animals with hard parts and a good geological record, such as clams or snails, is not observed. "Paleontological researches," Wallin (1927) noted, "have shown that new species have arisen suddenly. Furthermore, it seems that in certain geological eras a large number of new species arose simultaneously." With the 1972 paper by the pale-ontologists Niles Eldredge and Stephen Jay Gould, discontinuity in the fossil record became widely known as "punctuated equilibrium."

Although Lamarck, Darwin and Wallace, and a voluminous recent literature lack the details of the species-origin process, evidence is overwhelming that life forms change through time. There are different theories to explain this, but evolution is a fact. More than 200 years of worldwide scrutiny, observations, and measurements in nature and labo-ratories reveal no observed contradiction or anomaly.

Mendel Challenges Darwin's Heresy

Gregor Mendel (1822–1884), a Catholic priest who enjoyed an extensive correspondence with the pope on Darwin's heresies, was a superb scien-tist, educated at the University of Vienna. He discovered rules of trans-mission genetics. Mendel strove to refute Darwin's idea of descent with modification, which flatly contradicted Genesis. Mendel grew peas to show that crosses of differing purebred parents (hybridization) produced offspring with reversible "blended inheritance." Backcrosses of hybrids with purebred parents generated a return to fully unblended parental traits. Thus, Darwin was wrong. Species were static. Darwin's "inherited change," short-lived and reversible, certainly did not form new species. Mendel's paper "Experiments on Plant Hybridization" was read at a meeting of the Natural History Society of Brünn in Moravia in 1865, but was little understood. Mendel's inheritance rules were ignored and forgotten until their rediscovery in 1900.

The Neo-Darwinist Synthesis

The stasis of Mendelian genetics is inconsistent with Darwin's, Wallace's, and Lamarck's ideas of evolution as change. In the first half of the twentieth century, attempts to reconcile Mendel's stasis with Darwin's, Wallace's, and Lamarck's evolution led to the "new synthesis" (the "modern synthesis," or "neo-Darwinist theory"). This algebraic merger of Darwin's gradualism with Mendel's stable heritable factors (fixity, genetic determinism) rejected Darwin's and Lamarck's idea of inheritance of acquired variation. Small changes in Mendelian factors, later called genes, replaced pangenesis. The accumulation of random mutations in DNA was touted as leading to speciation and taught as fact throughout the twentieth century. Even now "the individual" is declared a unit of natural selection, and evolution is often defined as change in gene frequency in natural populations through time.

The notion persists that Mendel's factors, the genes described in molecular terms as stretches of nucleotide sequences in DNA, provide the key to life's mysteries. Neo-Darwinists tend to ignore any ecological concept of community and deny the idea that communities can be identified as units that evolve. Many zoologists and evolutionary biologists categorically deny the existence of "group selection." Dawkins (1990) promulgates the idea of "the selfish gene." But genes by themselves are only parts, not individuals. The selfish gene is an imaginative literary device; however, since genes (chemical sequences in DNA molecules) are never selves, it is as misleading as Spaceship Earth.

Russian "symbiogeneticitsts" rejected the explanation of evolution by natural selection that Wallace and Darwin presented in 1858 for changes in life forms through time. How, the Russians asked, could an elimination process explain the generation of life's diversity?

All organisms are composed of cells: either bacterial (prokaryotic) cells or communities of bacteria that fused to form cells with nuclei (eukaryotes). This concept was introduced by Kozo-Polyansky, a young Russian botanist unknown to the English-speaking world. That mitochondria (the membrane-bounded bacteria-sized bodies found inside most eukaryotic cells) and chloroplasts (the photosynthetic organelles, usually green, found inside all algal and plant cells) originated as free-living oxygen-respiring and oxygenic photosynthetic bacteria, respectively, was understood worldwide by the late 1970s (Margulis 1998). One great transition in the history of life occurred as toxic oxygen (O_2), the waste gas of cyanobacteria, threatened the predominantly anaerobic life forms that abounded in the Archean eon. One solution was the symbiotic acquisition of free-living bacteria that could respire oxygen in addition to tolerating anoxia; they became mitochondria. Phagocytotic acquisition of oxygen-using bacteria and their symbiogenetic retention catapulted early eukaryotes into increasingly oxygen-rich environments. Oxygen respiration provided them with a boost in energy relative to their other options (fermentation, nitrate respiration). Simultaneously, these protoctists gained access to vast volumes of aquatic habitat that had poisoned their oxygen-intolerant ancestors. Ambient oxygen continued to increase in concentration, and now O_2 gas constitutes nearly 21 percent of Earth's atmosphere.

Reciprocity, Redundancy, and Competition

Living systems range in size from Gaia down through biomes, ecosystems, biocoenoses (communities of organisms), individual multicellular organisms, and single-celled eukaryotic organisms to the tiniest bacterial cells. The biosphere, simply the place where life exists, is not equivalent to Gaia. The planetary Gaian system functions as interrelated dynamic life processes. Gaia's system components reciprocally modify habitats everywhere in the biosphere. The physical aspects of the biosphere change in ways correlated with evolution of life. Nucleated cells are co-evolved microbial communities, tightly and long-integrated energy-expending, matter-cycling components of Gaia. (See plate IV.) Life and its many habitats are ultimately linked, as Ian McHarg (2006) insists: "The Earth is not divisible." The "biosphere" is the 20-kilometer-deep hollow sphere of water, air, and land from above the tallest mountain to some meters below the ocean abyss. The biosphere, the place inhabited by life, is often depicted on maps as "biomes" (e.g., northern tundra, northern boreal forest, temperate deciduous forest, equatorial tropical rain forests)—large geographically and ecologically identifiable territories whose component ecosystems are best understood as portions of the biosphere large and diverse enough to entirely cycle chemical elements of life: hydrogen, carbon, nitrogen, oxygen, sulfur, phosphorus, and the more minor minerals required for continuity of all life. Elements cycle more rapidly within ecosystems than between them. This tends to make ecosystems robust and stable. Ecosystem diversity, cyclicity, and variation within communities and populations supply redundancy for life's continuation on this materially closed, energetically open, incessantly changing Earth. I view differential population growth due to measurable reproduction rates, competition, and natural selection as Gaian "quality control." Reciprocity is an emergent property of Gaia: life in and between interactive communities shapes the Gaian system, and natural selection shapes and limits the intrinsic tendency of expansion of cells, individuals, communities, and ecosystems. Changes in the environment during growth and development are also known to result in phenotypic variations in populations of organisms (Blumberg 2009).

Nested Communities, Multicellularity, and Community Ecology

Gaia and symbiogenesis are profound ecological ideas that recognize "community" as an essential unit of nature (figure 8.1). Social relations are those between members of the same populations of the same species, as was discussed in chapter 7. Different species or types of organisms, those that live in the same place at the same time, constitute communities in the Gaian planet-wide ecosystem. Communities, whether aquatic (pond, marsh, swamp, hot spring) or terrestrial (soil, tundra, taiga, alpine, meadow, desert), shelter many more organisms than are active at a given moment. Spores, cysts, larvae, seeds, statoblasts, tuns, fruits, and many other dormant stages reside within them. Such propagules represent potential for later population growth when seasons or other conditions change (Brown et al. 1985).

Gaia is composed of nested communities with subsystems from biomes to microbial associations. Microbial communities range from loose and temporary partnerships of lichen-like amalgams (e.g., *Geosiphon pyriforme*) to permanent integration by symbiogenesis that formed motile, mitochondriate (oxygen-respiring) ancestral eukaryotes, including humans. (See plate VIII.)

Bacterial communities may help us understand both inherited variation and animal and plant cell and tissue differentiation. Ivan Wallin (1883–1969), the brilliant American "symbionticist," did not know the

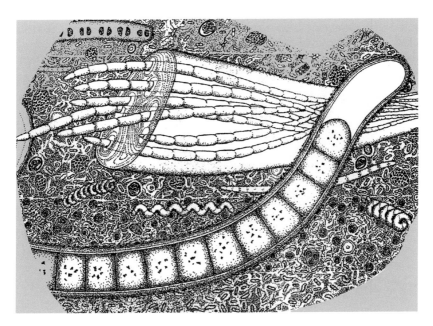

Figure 8.1
The cyanobacteria *Microcoleus cthonoplastes*, *Lyngbya aestuaria*, *Johannesbaptista* sp., and *Spirulina* sp. dominate the photosynthetic microbial mat community at Matanzas, Cuba.

Russian literature. He recognized, independently, that eukaryotic individuals—protoctists, fungi, animals, and plants—are composite (Margulis and Chapman 2010). "Community ecology" generates variation in populations that we view as differentiation of cells and growth and development of individual organisms. Wallin recognized that development biology is a kind of community ecological succession. The following two quotations from his 1927 book *Symbionticism and the Origin of Species* hint at this great insight, an insight ignored by some and denigrated by others:

Did these variations of bacteria that gave rise to the protozoa and algae [both protoctists] originate as the result of environmental influences, or are they the result of some other factor? We have maintained, and the evidence submitted by many observers supports the contention, that bacteria are changed or modified by environmental influences.

Bacteria in nature are known to produce the same enzymes and metabolic products that are formed by the tissues of plants and animals. Bacteria have been shown to develop intimate symbiotic relationships with plants and animals. In these symbioses it has been observed that 'new' cells develop (modifications of preexisting cells) to harbor the symbionts. Examples have been given in which new organs developed in response to the presence of microsymbionts. Are these observations to be interpreted as biological digressions without any fundamental significance? . . . We are forced to the conclusion that mitochondria constitute a fundamental factor in cellular differentiation.

Darwin's admonition to study "oddities and imperfections" applies to most protoctists. They include more than thirty phyla of seaweeds and other algae, slime molds, foraminifera, ciliates, and amoebae). Many such "odd" and "imperfect" life forms are available for study. Many are extant; some, such as ebridians, are known primarily as fossils. Lynn Margulis and about fifty colleagues heeded Darwin when they wrote the *Handbook of Protoctista* (Margulis and Chapman 1990, 2012). The second edition describes these peculiar organisms, all clearly evolved by symbiogenesis, that help us trace the evolutionary history of nucleated cells.

Seeing Trees

Study of Gaia and symbiogenesis requires knowledge that transcends traditional preconceptions, dogmas, and the boundaries of individual academic disciplines. An approach both interdisciplinary (among fields of science) and transdisciplinary (beyond any specific scientific field) is needed. Though ecologists may study communities, rarely do they see what Vladimir Vernadsky (1945) described as "the outline of a single and complete system in the apparent chaotic appearance of nature." I suggest that the proponents of current mainstream views of Earth (geologists) and life (biologists) seem unaware that the concept of community is fundamental and requires simultaneous geologic and biologic observations. Communities of different kinds of life living together are systematically ignored. Whereas most biologists see no problem with the current assumption of the "individual as unit of selection worldview," most geologists ignore the "evolutionary mechanism" debate. If both of the ecological-evolutionary concepts—Gaia and symbiogenesis—are supported by strong evidence, why does neo-Darwinism still dominate most evolutionary thought?

Missing Seven-Eighths of the Story

Most proponents of neo-Darwinism are zoologists, even though the genetics (Mendelian), the life history (diploids with gametic meiosis), and the metabolism (chemoorganoheterotrophic) of animals are the least varied of the five most inclusive taxa. The zoological view that dominates the literature on evolution is short-sighted. Stephen Jay Gould—paleontologist, evolutionary biologist, and historian of science—once characterized the Precambrian as a time in which "nothing happened for ever so long." "On the contrary," the evolutionist Lynn Margulis replied (1998), "the Precambrian, [Archean and Proterozoic eons] is when everything important in the evolution of life happened."

Indeed, most of the sensory, locomotive, and metabolic pathways, community organization reproductive strategies, and distinctive life cycles that exist today originated and evolved during the first 3,000 million years of Earth's history. The fossil record of bacteria, the earliest life, extends back in time to the first sedimentary rock record. Protoctists, permanently integrated communities of bacteria, first appeared more than 1,200 million years ago. Many strange species of the soft-bodied sand-dwelling life forms known as Ediacarans evolved no more than 700 million years ago. They flourished and went extinct before the appearance of most animal phyla. Zoologists and paleontologists study the fossil record that begins with the sudden appearance of animals with hard parts 542 million years ago at the base of the Cambrian. Plants and fungi, primarily land dwellers, evolved relatively late, roughly 450 million years ago. These organisms, always found in structured communities as ecosystem components, altered the Earth that they and we now inhabit.

Mutation

Neo-Darwinists champion the idea that new taxa (species, genera, families), their organs, tissues, organelles, and phenotypic traits evolve by direct filiation—the accumulation of random mutations that underlies diversification from totipotent ancestors. The term "group selection" is anathema to neo-Darwinists because it flies in the face of mathematical models that assert that selection occurs on the individual level or the gene level. Group selection is intrinsically at odds with the exclusive selection of the individual, with its anthropocentric notions of selfishness and altruism.

The concept of "individual evolution by direct filiation" was modified in the 1980s by compelling evidence that showed that mitochondria and chloroplasts were acquired by symbiogenesis. These prepackaged live chemical wizards inserted oxygen respiration and photosynthetic capability into lineages of organisms incapable of such powerful metabolic modes. The formerly ridiculed idea of symbiogenesis "mutated" to become a new self-evident truth. The acceptance of Mereschkhovsky's idea of chloroplast symbiogenesis (1905) and of Wallin's later advocacy of a similar bacterial origin of mitochondria by all serious scientists brought into the mainstream the idea that symbiogenesis occasionally was essential to the evolution of modern life (Cavalier-Smith 2003).

There is evidence, some of it from molecular biology, that eukaryotic cells descended from two different types of bacterial ancestors: archaebacteria and eubacteria. Symbiogenesis was, therefore, responsible for three major evolutionary events in life's history: the permanent integration of an archaebacterial-eubacterial community into the first eukaryotic cell; the acquisition of oxygen-respiring mitochondria by the ancestors of all animals, fungi, and plants; and the acquisition of photosynthesizing chloroplasts by the ancestors of algae and land plants. Direct filiation by accumulation of random DNA base pair mutations has not been shown to have led to even one evolutionary change of comparable magnitude, yet in 2003 Thomas Cavalier-Smith, who denies that symbiogenesis is "a giant favorable mutation" or a "macromutation," still contended that "mutation is the greatest innovator by far."

Minor Irritations

Neo-Darwinists, whose ideas of evolution disregard the environment in general, geological, astronomical, and other temporal cycles (see chapter 9), and communities, often overemphasize "competition" and "struggle" between individuals. They assume the superiority of animal multicellularity, as did Darwin (1859): "Thus, from the war of nature, from famine and death, the most exalted object of which we are capable of conceiving, namely the production of the higher animals directly follows."

But some members in all of the most inclusive taxa (Kingdoms: Prokaryotae, Protoctista, Fungi, Animalia, and Plantae) have colonial and multicellular descendants. Multicellular bacterial fossils are found in thin sections of chert from the Gunflint iron formation in western Ontario, which is 2,000 million years old. Darwin himself—a naturalist—wavered in his anthropocentrism. In a letter to the botanist Joseph Hooker, he advised against use of the terms "higher" and "more evolved" on the ground that all life currently in existence had survived the "struggle for existence." Extant organisms have evolutionary histories that date from the origin of life. None is more evolved or higher. More recently evolved taxa, including humans, seem to have expanded the volume they occupy and increased their energy efficiency relative to the earlier, less volumetrically extensive ecosystems. But this expansion and increase in energy flow is a property of Gaia's components, not of the evolution of Darwin's "exalted objects" (English-speaking humans).

Cyanobacteria and other autotrophic bacteria that produce their food and their bodies from carbon dioxide by the use of external energy sources (light or the oxidation of inorganic chemicals) are Earth's primary producers. The rest of life depends on them. Bacteria, considered in their entirety, are arguably the most capable, most self-sufficient, and hardiest organisms on the planet. Certainly they are the earliest and the most persistent. They still make up most of the mass of life. They include extremophiles that thrive at temperatures, under pressures, and in anoxic conditions that define the limits of the biosphere. We humans have more than 2,000 kinds of normal bacterial associates. Skin and intestinal bacteria accompanied cosmonauts into space. The lunar astronauts did not visit the surface of the Moon alone; they landed in the company of their bacterial symbionts.

Missing the Forest

If Gaia and symbiogenesis are supported by evidence, why are they ignored, denied, or denigrated? Why are these examples of symbiotic partnerships, populations, communities, and ecosystems as naturally selected units unrecognized? What establishes zoologists as the experts on evolution? Zoovolutionists, it seems to me, speak far outside their field. Do factors besides evidence influence the disregard (or acceptance) of symbiogenesis and Gaia?

Ian McHarg (1920–2001), a pioneer of the environmental movement and one of the fathers of Geographic Information Systems (GIS), described the effects of specialization of scientific disciplines as it pertained to his field of regional planning and landscape architecture as follows:

The whole system [region] is one whole system, only divided by language and by science. Our job is to reconstitute the region and all its processes again, like putting together Humpty Dumpty.

Humpty Dumpty sat on a wall,
Humpty Dumpty had a great fall,
And all the King's horses and all the King's men
Couldn't put Humpty together again.

This is what modern science is; the egg is shattered, all the fragments lie scattered on the ground. The fragments are called geology and physics and chemistry and hydrology and soil science, plant ecology, animal ecology, molecular biology, and political science. There is no one who can put together again the entire system. Information fragmented is of no use to anybody. What we always need to proceed is really the one whole system . . . somebody has to put it together again.

Science and university departments are devices by which integration is all but impossible. As we know, academics resist integration with incredible ferocity. Because of the requirement scientists in universities have for purity and success, clearly the need to integrate is resisted by scientists. Holism therefore is extremely difficult. . . . And, so, we set the scientist this very difficult task for which they are remarkably untrained. We ask them to group together all these independent spectral views of the universe into one whole system. It is very difficult, but once one has it, one has the best description natural science can give us . . . as a single interacting process understood in the context of its long past. (McHarg 2006)

Apartheid or Essence

The scientific ideas of Gaia and symbiogenesis provide a desperately needed community view of life and of planet Earth, but they are unorthodox concepts. Many laymen and scientists believe that science transcends the political-economic-social fray and assume that scientists just collect facts. This unquestioned assumption is a cognitive trap. Science is an intensely social human activity. Scientists are initiated into academic fields ("thought collectives") and develop confidence for the accepted "facts," which may be precisely but inaccurately measured and modeled (Fleck 1935). Ideas are often accepted with parsimony interpreted as evidence; such "misplaced concreteness" pervades scientific research. Unexpected and astonishing, Gaia and symbiogenesis contradict accepted facts and invite serious scientists, historians, and scholars to ask themselves "What would I maintain is the essence of science?" and "Am I open to surprise?" McHarg was correct: the world is whole, but various disciplines divide it up into parts that lead us to lose sight of systems such as Gaia. "Academic apartheid" (Lovelock's term, personal communication) is a colossal obstacle to the integration of knowledge required to sustain a healthy human future. Since no species of mammal has survived longer than about 10 million years, I assume the inevitability of *Homo sapiens'* extinction. We must anticipate a greatly reduced human population. I suspect that our extinction will be more pleasant if we recognize the extent to which we are embedded in nature on planet Earth.

III
Earth

9

Cosmic Rhythms of Life

Bruce Scofield

Here, recognizing the pervasive importance of timing in all the activities of life, Scofield reviews the various levels of interrelationships between cosmic and biological cycles.

From the origin of life as bacteria to the sexual cycles of plants and fungi, the activities of all organisms occur in the constantly changing temporal environment created by the geophysical and astronomical cycles of Earth, the Moon, and the Sun. Life has sensed and responded to this environment for perhaps 3,800 million years and has used these geophysical and astronomical influences as a structural framework. Life's processes respond to environmental constraints (Scofield and Margulis 2011). Sensitivity to temporal cycles augments survival and reproduction. Biological rhythms based on geo-celestial timing alert the living to changes in the environment. They cue the senses and memories that optimize environmental navigation (Foster and Kreitzman 2004). Responses help locomotion, feeding, mating, spore and seed germination, flowering, and other specific aspects of survival. Sensory behaviors, physiologies, and genetic repertoires that register cycles (e.g., days and seasons, tidal and other lunar rhythms, Earth's magnetic field, and the solar cycle) have evolved, as recognized since antiquity (Scofield 2004).

The primary geophysical periodicity to which nearly all organisms respond is the rotation of Earth relative to the Sun; i.e., the solar day of 24 hours. A related period is the tidal or lunar day of 24.8 hours, which denotes the rotation of Earth relative to the position of the Moon. During the course of one 24-hour rotation of Earth, the Moon advances in its orbit by about 13 degrees of celestial longitude, requiring an additional 0.8 hour of Earth rotation to complete. Another lunar period is the synodic cycle, or the cycle from new Moon to new Moon, of about 29.5 days. Half of this—14.77 days, spanning the time from new to full

Moon—is called the semilunar period. The solar year of 365.24 days, which marks one orbit of Earth around the Sun, is a period during which the ratio of light and darkness predictably varies in a cycle that creates the seasons. This annual cycle is most pronounced at high latitudes. Cycles longer than a year that are driven by solar activity exist, such as the 1.3-year cycle of the solar wind, the Schwabe sunspot cycle of 11.1 years, and the Hale cycle of 22 years.

Internal rhythms are fundamental features of life. "Biological clocks," diurnal and seasonal patterns of enzyme activity, flowering and fruiting, reversible dormancy of spores seeds and cysts, and many other phenomena that match environmental periods are a ubiquitous characteristic of life. Yet there are wide variations between species and even between individuals in their cyclical responses. In some species, one clock, or clock system, controls all life functions; in some species, multiple clocks may be coupled; and in some species, "master" clocks drive "slave" oscillators. In nature, absence or malfunctioning of physiological timing may be lethal.

Biological rhythms were recognized by early agriculturalists and natural philosophers. In his description of the Mediterranean sea urchin, Aristotle mentions that the size of its ovaries varies according to the lunar cycle. Modern study of biological rhythms began in 1729, when the French astronomer Jean Jacques d'Ortous de Mairan observed the persistence of the cycle of daily leaf movements of a heliotrope species maintained in continual darkness. Charles Darwin published comparable observations in his 1880 book *The Power of Movement in Plants*. Research on efficient tobacco propagation led to the discovery in 1920 that a plant flowering at specific times in the year was influenced by the cycle of the day/night (light/dark) ratio within the annual cycle. This property is called photoperiodism. The German botanist Edwin Bünning argued in the mid 1930s that the mechanisms behind both seasonal flowering and diurnal leaf movement rhythms was a roughly 24-hour endogenous, or internal, "biological clock."

The underlying nature of biological rhythm physiology has been debated since the beginning of the nineteenth century. In opposition to Bünning, an argument was made that the "clock" is a sensory response to any number of direct environmental signals, exogenous and driven by an external *Zeitgeber* (time giver). Laboratory data challenged this view, and studies conducted at the South Pole with hamsters, *Drosophila* sp. (fruit flies), and *Neurospora* sp. (pink bread mold) and in space with *Neurospora* sp. pointed to an endogenous self-sustaining oscillator that can run without any obvious environmental cues. Biological rhythms are mostly independent of the environment. They are driven by complex internal molecular processes and not by direct external forcing. But these rhythms are entrained by environmental cues; external signals are essential for their operation in nature.

Earth's Rotation and Circadian Rhythms

The alternation of the light-dark cycle, the geophysical period due to rotation of our planet, is the most important environmental signal for life near Earth's surface. This signal is experienced as a repeating pattern of light and dark that occurs as Earth faces toward or away from the Sun during the course of a solar day. Biological rhythms of approximately 24 hours have been found to be a general feature of the physiology of all organisms studied: prokaryotes (bacteria, including archaebacteria) and eukaryotes (amoebae, algae, ciliates, seaweeds, slime molds, etc.), plants, fungi, and animals. (See appendix A.) These extensively studied rhythms are referred to as circadian (circa = about, dian = day) (Moore-Ede et al. 1982).

Authentic circadian rhythms are defined by three main characteristics. The first is the persistence of an approximately 24-cycle in constant light or dark conditions. The persistence of rhythm and deviation of this cycle from precisely 24 hours, the period, is species specific. The rhythms of organisms may continue without external signals for only several complete cycles; others are able to "free-run" for weeks or months in the absence of cues. Second, circadian rhythms persist throughout a wide range of temperature. Thus a stable rhythm independent of ambient temperature is maintained throughout the day or the year. Third, circadian rhythms can be entrained by light/dark (L:D) cues. They can also be entrained by rapid temperature changes of >10°C, by food availability, by social cues, by electromagnetic field strength, by atmospheric pressure, and by other environmental signals.

In circadian rhythms a phase relationship is established between subjective day and external day, or night in nocturnal organisms, by response to an external environmental signal. In some organisms the phase of their circadian period is set by the onset of light, in others by the onset of darkness. Circadian rhythms can be modulated; i.e., phase shifted, by the presence of specific environmental signals, such as a strong light pulse at a point other than the expected onset of light or darkness in the diurnal cycle. This ability to "phase shift" accounts for the ability to respond to the changing day/night ratio of the seasons.

The study of biological rhythms has accelerated in recent years. Although circadian rhythms were once thought to be present only in eukaryotes, they were discovered in cyanobacteria in the late 1980s. A decade later a molecular model for "the circadian oscillator" was developed. In it a photoreceptor senses alterations of light and dark and the sensed information activates an oscillating system within the cell that is essentially a negative feedback loop for protein synthesis. This oscillator is capable of being reset by the photoreceptor when new photic information reaches it at critical points in its cycling. Chemical messengers then relay this "time-data" to other parts of the organism where it serves a variety of regulatory functions (figure 9.1).

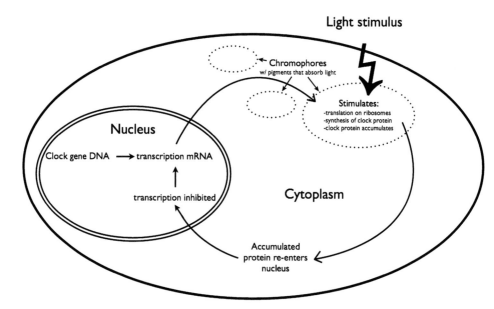

Figure 9.1

A generalized model of the working of a biological clock: a eukaryotic, negative feedback loop, circadian system. A specific chromophore, usually a small protein-associated pigment sensitive to the light signal, accumulates in the cytoplasm. When the protein level reaches a threshold, it re-enters the nucleus, binds to the promoter region of the clock gene, and inhibits transcription of the clock gene product which causes production to diminish. The lowered protein concentration of the clock gene product then leads to renewed expression of the clock gene. Entrainment to the light signal produces a circadian cycle that matches the environment. Note that circadian systems employ multiple clock genes.

Circadian Rhythms in Bacteria

Since circadian systems were observed to control the cell-division cycle, and prokaryotes divide more frequently than once per day, the assumption prevailed that bacteria must lack circadian cycles. It was thought that division cycles of fewer than 24 hours would uncouple any circadian clock, yet the detection of a circadian rhythm in the common, abundant marine cyanobacterium *Synechococcus* sp. reveals this assumption to be incorrect.

The separation of anaerobic nitrogen fixation from oxygenic photosynthesis is a challenge for photosynthetic organisms that have both of these metabolic activities. In filamentous cyanobacteria, specialized cells called heterocysts spatially segregate nitrogen-fixation enzymes that are poisoned by oxygen. *Synechococcus* sp., a coccoid unicell, separates oxygenesis from nitrogen fixation in by time. Using its circadian cycle, which also regulates abundance changes in mRNA, amino-acid uptake, protein synthesis, light/dark entrainment, and components of the cell-division cycle, *Synechococcus* photosynthesizes by day and fixes nitrogen after dark. (See figure 9.2.)

Genes specific to control of circadian timing in *Synechococcus* sp. were isolated in the early 1990s. It was found that a cluster of three clock-specific genes, kaiA, B, and C (*kai* = cycle in Japanese), were necessary in combination for persistence of the daily cycle, and that the deletion of any of them caused arrhythmicity. The kai genes are unique to prokaryotes, although two are homologous to gene sequences in

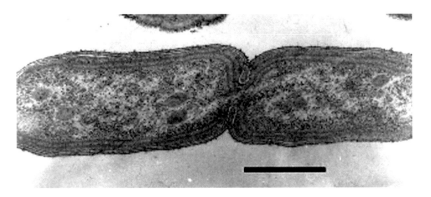

Figure 9.2
An oxygenic, photosynthetic cell in division: cyanobacterium *Synochococcus* sp., in which biological clocks have been studied. Scale bar = 0.5 μm.

archaebacteria. However, the kai genes are not related to eukaryotic clock genes such as those found in *Drosophila* sp. (fruit fly), *Neurospora* sp. (bread mold), or *Mus* (mouse).

Molecular evidence suggests that the cyanobacterial circadian cycle is extremely ancient. KaiC, the oldest of the three, may have evolved in cyanobacterial ancestors as early as 3,800 million years ago. The kaiBC genes are estimated to be 3,500 million and 2,300 million years old, perhaps related to the increase of oxygen concentration in the atmosphere. Gene kaiB may have been laterally transferred from cyanobacteria to other eubacteria and to archaebacteria. That the kaiA gene is limited to cyanobacteria suggests a more recent origin. It is thought that the cyanobacterial kaiABC cluster probably was in place 1,000 million years ago.

Circadian Rhythms in Protoctista

Circadian rhythms are detected in many single-cell protoctists (algal protists such as diatoms, *Acetabularia* sp., *Chlamydomonas* sp., *Euglena* sp., *Gonyaulax polyedra*). An excellent example of the versatility of circadian rhythms is found in *G. polyedra*. In this dinomastigote alga, a bioluminescence circadian rhythm and daily cycles of photosynthesis and cell division are entrainable by the natural day/night sequence or by single short light pulses. Pulses of light phase shift the rhythm if applied at specific times. Cell division occurs at the end of the night phases; light cues then reset the clock. Mitotic cell division does not control the clock; rather, the clock controls the cell cycle. The circadian rhythm of photosynthesis responds to maximum light at midday, but entrainment sensitivity occurs at dawn—that is, at the start of the light phase. *Gonyaulax* sp. cells display at least two separate oscillators, each receptive to different wavelengths of light. Its rhythm of bioluminescence is modulated by absorption at blue wavelengths, whereas aggregation behavior rhythms respond to red wavelengths (Sweeney 1987).

The circadian clock even drives mating-type cycles in *Paramecium micronucleatum*, a ciliate. Sexual processes in *Paramecium* sp. are of three types: autogamy (self-fertilization by nuclear fusion without division), binary fission (asexual karyokinesis and cytokinesis), and conjugation (mating between complementary multiple mating types without division). Gender changes within individuals occur on an entrainable circadian rhythm. This peculiar sexual reactivity of *P. micronucleatum* was reported in 1966 by Audrey Barnett, a Ph.D. student of Tracy

Sonneborn. She found that the mating type rhythm is endogenous. Cells change gender on a daily (circadian) rhythm. They can be entrained even during division. Every ciliate, including *P. micronucleatum*, contains nuclei of two kinds: the sexual micronucleus and the physiological macronucleus. The ability to cycle is transmitted by a gene in the micronucleus, whereas the control of the circadian rhythm is expressed by macronuclear RNA synthesis.

Acetabularia sp., a large single-cell green alga with thousands of mitochondria and chloroplasts, is 5 centimeters high and has a "cap" about 1 cm in diameter. Generally found in shallow, sheltered marine waters, it attaches to rocks and other shallow substrates present in or near tropical coral reefs. The location of the circadian oscillator cannot be simply located in a nuclear gene, since removal of the single basal nucleus fails to terminate circadian rhythm. The circadian rhythm persists for weeks in the absence of the nucleus. The nucleus may influence the daily rhythm, but exactly how is not known. Further, a "circaseptan" rhythm of about one week was observed in the amplitude (strength) of the fundamental circadian rhythm. The paramecium and algal examples illustrate the difficulty of locating the circadian mechanism, a problem that is not completely resolved by any molecular model.

Circadian Rhythms in Other Eukaryotes

The circadian rhythms in most if not all eukaryotes operate in similar ways and serve multiple important functions. Studies of laboratory organisms suggest that all eukaryotes use the same or similar molecules to control their circadian rhythms within the transcription-translation oscillator model. As translation of DNA to messenger RNA proceeds, the protein gene product of specific genes accumulates in the cytoplasm. The accumulation to a certain concentration inhibits further transcription of its own genes. When the gene product declines to a given level, translation again begins. The circadian oscillator thus uses negative feedback loops to sustain itself. Although this circadian oscillator-protein-feedback model appears to be common to both eukaryotes and prokaryotes, the specific proteins involved differ. The molecular circadian system is polyphyletic; it evolved independently more than once.

Animal circadian rhythms are entrained primarily by photoreception through the retina. Extraretinal photoreceptors in the brain or connected nervous tissue may supplement light sensitivity of the eyes, however. The anatomical location of the circadian oscillator in animals varies widely;

in eyes of gastropods, in eyestalks of some crustaceans, in insect brains, or abdominal ganglia, or brains of other organisms. In reptiles and birds the oscillator is located in the pineal gland, whereas in mammals it is in the hypothalamus. Circadian biological clocks in animals share common ancestry; mammals and insects have in common at least some of the same clock components.

In vertebrates, the central pacemaker is the suprachiasmatic nucleus (SCN), located in the hypothalamus. The SCN produces circadian rhythms by means of gene-product negative feedback loops in specialized cells, which then relay information via neural and endocrine pathways to the rest of the organism, including other peripheral clocks. The SCN controls the rhythmic release of melatonin, which carries the rhythmic signal via the bloodstream to the rest of the body. The processing of light by vertebrates for the circadian oscillator differs significantly from the processing of light for vision and appears to involve brightness (photon count) receptors that are separate from rods and cones. For example, the mammalian eye has parallel pathways for vision and brightness; the latter has dedicated photoreceptor cells that comprise about 1 percent of the total ganglion cells in the retina (see chapter 20).

Photoperiodism

The seasonal progression of day length is related to Earth's orbit around the Sun. Day length is modulated over the course of the year by the tilt of Earth's axis: the day:night ratio is 12:12 at the equator but increases to 0:24 at latitudes above 66 degrees. The annual cycle's changing day length challenges animals and plants, particularly at higher latitudes, and influences feeding, reproduction, growth, molt, migration, and hibernation behaviors. Animals continuously entrain their circadian oscillator by constant phase resetting, though some long-lived species have been shown to possess an endogenous free-running (environment-independent) period of 12 months. In birds, the photoperiodism response is more robust and predictable in long-distance migrators than in species that live near the tropics, where the strength of annual light cues is limited.

Plants keep time. They clearly measure day length, but the multi-gene molecular mechanism is not completely understood. The phase (i.e., day/night) and the period (duration of full cycle) of time keeping in plants probably are under separate genetic control.

Tidal and Lunar Rhythms

The second most conspicuous rhythmic cosmic signaler for life is the Moon. The gravitational pull of the Moon on Earth produces a daily tidal cycle of 24.8 hours; the two high tides are spaced about 12.4 hours apart. A high tide that occurs on the side of Earth that faces the Moon also occurs simultaneously on Earth's opposite side. Coastline shapes and land masses distort the timing of local daily tides, making tidal schedules highly irregular. The periodicities generated by the inexact and multifaceted lunar cycles in marine animals are not as precise as free-running circadian rhythms.

The synodic cycle of 29.5 days is defined by the number of days elapsed between alignments of the Moon and the Sun, i.e., new Moon (Sun-Moon-Earth) or full Moon (Sun-Earth-Moon). The combined gravitational forces of the Sun and the Moon at these two alignments correlate with the highest tides. Because of the proximity of the Moon, the Sun's gravitational contribution is only about 45 percent. The distance between Earth and the Moon varies by about 13 percent over a cycle of 27.5 days, called the "anomalistic month." At perigee (when Earth and the Moon are nearest one another), the Moon's gravitational force is stronger and the coincidence of perigee with a full or new Moon produces maximal high tides.

The Moon's influence on the environment is not limited to variations in nocturnal illumination. The atmosphere displays lunar tides as the upper layers are affected by the Moon's gravity and this modulates air pressure. Magnetic field data also show adjustments to the Moon's phases. Ocean tides produce complex environmental changes, and tidal triggers in an organism include changes in temperature changes, in turbulence, in water pressure, and in salinity. Few studies have compared the responses of organisms to the various cycles of the Moon, and the subject is, in many respects, in its infancy.

Tidal and lunar periodicities are prominent features in the lives of many marine organisms, though there have been few extensive studies. One of the most studied tidal organisms is the fiddler crab, *Uca* sp. (Brown et al. 1970; Palmer 1995). These arthropods live in burrows and feed as the tide ebbs and so are subjected to repetitive fluctuations of light, temperature, and tidal submergence. It has been found that they have a 24-hour circadian cycle of shell color—darker at dawn, lighter at dusk. However, their running activity retains a tidal periodicity of about 24.8 hours, 50 minutes later each day. Further, in *Uca pugilator* the reproductive rhythm is semilunar: larvae are released at new or full Moon at the hour of high tide. It appears that multiple clocks operate in this crab.

One of the more curious of the studies that have raised the endogenous/exogenous issue in lunar rhythms involved a study of oysters transported from Connecticut to Illinois. The oysters were kept in a pan of unagitated water, and the frequency of shell opening and closing (feeding behavior) was recorded. At first, maximum activity occurred at high tide for Connecticut, but after two weeks the rhythm phase shifted three hours later in lunar day, roughly equivalent to the tide in Illinois, where it then stabilized. One explanation proposed was that the organisms were responding to tides in the atmosphere that register as subtle changes in barometric pressure (Brown et al. 1970). Contrary to this finding, mussels collected in Massachusetts and taken to California maintained their East Coast tidal rhythm and did not adjust to West Coast tidal rhythms until they were exposed to these tides.

Lunar rhythms are also found in marine protoctista. Many species of planktic foraminifera have a reproductive cycle that is characterized by the alternation of two generations, haploid and diploid, with different modes of reproduction. At precise times, apparently synchronized to the lunar synodic cycle, free-swimming haploid gametes leave the shell (called the "test") of the foraminiferan, meet with others of their kind, and fuse. This begins the diploid phase of the life cycle. Gametes released in the ocean require consolidation in time and space. Precise synchronization is necessary to secure gametic fusion, fertile unions, and thus continuation of the species. In a study of three foraminiferan species, Bijma et al. (1990) found that each has its preferred time for gamete release relative to the full Moon (ibid.). Even more remarkable is the lunar-synchronized reproduction of at least 107 species of coral along 500 kilometers of the Great Barrier Reef in the western Pacific. Nearly all release gametes 3–6 days after the full moon in October or November, about 4 hours after sunset (Endres and Schad 1997).

The Geomagnetic Field and Magnetotaxis

Earth is shielded from a steady rain of charged particles and cosmic rays by its own self-generated magnetic field. The magnetic field not only establishes an Earth-encircling electro-magnetic dipole framework, but it also exhibits fluctuations and periodicities. Geomagnetic periodicities reflected in magnetic field variations include responses to Earth's rotation

(24 hours) and the lunar day (24.8 hours), the synodic month (29.5 days), solar rotation (about 27 days), the solar year (365 days), variation in the solar wind (1.3 year), and the sunspot cycle (11 years for the Schwabe cycle and 22 years for the Hale cycle, on average).

Many organisms have evolved receptors that allow them to use the magnetic field for both timing and navigation. In a process called "biomineralization," magnetotactic bacteria synthesize tiny magnets in their bodies. They use attraction to Earth's magnetic field to orient in the water column (figure 9.3) and therefore settle appropriately in the sediment. Animals known for their navigating abilities include pigeons (which have magnetic particles under their skulls), honeybees (which have magnetite in their abdomens), tuna, trout, blue marlins, green turtles, whales, dolphins, and possibly humans. The magnetic sensory systems probably evolved early in life's history. Completely separate from other sensory organelles and organs, they have increased in sensitivity over time.

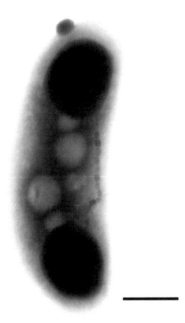

Figure 9.3
Negative-stain electron micrograph showing magnetosomes (sites of sensory detection of a magnetic field). The dark row of magnetosomes seen lined up inside a single magnetotactic bacterium are surrounded by the shadow of the bacterial cell. Scale bar = 0.5 μm.

Solar Cycles

Sunspots are visible evidence of both solar rotation and of a larger cycle of solar magnetism that affects Earth's magnetic field. The most recognized solar cycle is the Schwabe cycle of about 11 years, during which the number of sunspots increases and decreases. The Hale cycle of about 22 years accounts for the magnetic reversals that occur between Schwabe cycles.

Many correlations have been reported that link biological and solar activities. Well-documented rhythms of about 10 years, close to the Schwabe cycle, include crop yields, fish catches, and changes in the mammal populations of boreal forests. Insect populations show close correlations with the Schwabe cycle and thus are regarded as sensitive climate monitors. Populations of tent caterpillars predictably reach a maximum about two years before the peak of solar activity. Other insect populations appear to do the same. These population increases may be due to the increased solar irradiance at sunspot maximum. At least one unicellular alga (*Acetabularia* sp., whose circadian cycle was described above) shows a response to a solar cycle. Its rhythms are recorded in the geomagnetic Kp and aa indices and also in the solar magnetic field. In a database covering 14 years of *Acetabularia* sp. circadian cycle research, Franz Halberg, who reported the prominence of a 1.3-year cycle, also found in the rhythms of the solar wind.

Biological clocks that match environmental periods are a nearly ubiquitous characteristic of life, but there are wide variations between species and even between individuals. Their primary function is to regulate temporally the various cellular processes and behavioral components of an organism in its environment. "Biological clocks" are assumed to confer fitness, though only a few laboratory studies have demonstrated this. A lack of rhythm is thought to cause an organism to be active at the wrong times of day. Errors in schedules for mating or feeding are assumed to reduce the number of offspring.

The current molecular model of the circadian oscillator has established an endogenous source for circadian rhythm. Exceptions exist, and similar models for lunar rhythms have yet to be described. The endogenous oscillator system facilitates tracking of environmental signals, accurate adjustments (phase shifting), and hence increased fitness. But endogenous oscillators require sensory input and external triggers to set the phase. Although light is the primary trigger for the circadian system, any number of ways sensory input may be involved in phase setting can be imagined. Other periodic environmental entrainment signals for rhythms are food availability, social cues, hydrostatic pressure, agitation, and temperature change. Magnetic field variation and the solar wind may also trigger or modify endogenous rhythms. Life has deep links to a broader definition of the environment than is generally acknowledged. The study of biological rhythms helps record how life has responded to and internalized its environment by use of the natural geophysical and astronomical periods. The temporal guide aids in organizing the complex processes that life and its environment have co-evolved.

10

Life's Tectonics

Paul D. Lowman Jr. and Nathan Currier

The cycling of living systems may extend to continental drift and the plate tectonics that drive it. Here Lowman and Currier suggest that water-based cycling life extends to the water-based cycling of Earth's huge lithospheric plates, the raised portions of which we recognize as the continents.

In 1924, Alfred Wegener, a German meteorologist, published a revolutionary theory in a book titled *The Origin of Continents and Oceans*. This theory, or concept, soon became better known as "continental drift." It triggered several years of intense debate among geologists, most of whom, including Walter Bucher, an eminent Columbia University geologist, rejected it (Bucher 1933). However, many years later Wegener's theory was resuscitated as part of the theory of plate tectonics.

The essentials of plate tectonic theory, widely known, so they can be are summarized in the maps reproduced in this volume as color plates V and VI. These maps show Earth's "present" tectonic and volcanic activity (i.e., that of the past million years). This period is long enough to give a realistic picture of Earth's internal activity, but short enough that geomorphic features formed by this activity, such as volcanoes and rift valleys, are still recognizable despite erosion. The maps show the essential features of plate tectonic theory.

The plates themselves are large rigid segments of the lithosphere, bounded by some combination of sea-floor spreading centers, subduction zones, and transform faults. How many plates are there? A widely used mathematical model assumes twelve, but the number will be much larger if we consider many small crustal blocks to be plates. The best evidence for plate tectonics is found in the ocean basins, especially in the Pacific; the theory of plate tectonics began at sea. One of its main developers was Admiral Harry Hess, who kept his ship's echo sounder on during many trips across the Pacific during World War II. A marine geologist, Robert Dietz, coined the term "sea-floor spreading" and applied it to many phenomena. Space geodesy, with artificial satellites and radio astronomy used to carry out extremely precise long-distance measurements, has confirmed plate rigidity and motion directly by three independent methods: satellite laser ranging, radio telescope interferometry, and most recently GPS measurements. These features can be considered conclusively demonstrated. They imply the third element of the theory: subduction, the underthrusting of basaltic oceanic crust beneath the lighter continental crust. For example, the Pacific crust is currently being thrust beneath the South American plate.

Gaia's Extent

What is the relationship of Gaia (see chapter 8) to the rest of planet Earth (Westbroek 1991)? When did Gaia first detectably appear on this third planet of the solar system? To what aspects of the physical-chemical conditions on Earth do all living components of Gaia react? Which aspects do they ignore?

As summarized by Lynn Margulis (1998), Gaia is Earth's biotic-planetary regulatory system, which produces and removes gases, ions, metals, and organic compounds through the metabolism, growth, and reproduction of an estimated 30 million species of microbiota (bacteria, smaller protoctista and fungi), flora (organisms including algae, such as green seaweed and kelp, mushrooms, and other large fungi erroneously labeled "plants"), and fauna (marine larvae, insects, worms, even deep-sea pogonophorans and many other groups traditionally called animals). These interactions in aqueous solution lead to modulation of Earth's surface temperature, of acidity and alkalinity, and of the chemically reactive gases of the atmosphere and the hydrosphere (Schwartzman 1999). In recent years the Gaia hypothesis has been extended to assert that such biogenic climatic control underlies our planet's retention of water (Harding and Margulis 2009). Further, in contrast with Venus and Mars, H_2O, liquid H_2O. This literally vital liquid is not yet found in abundance elsewhere in the solar system (Sagan 2007). Water solute concentrations are modulated and regulated by life, as is indicated by the well-established example of the nitrogen-to-phosphorus ratio of the oceans. The aim of this chapter is to demonstrate that life's control of the oceans has, in turn, been of crucial importance to Earth's tectonic activity.

Comparative Planetology

James Lovelock's question "How has Earth kept its oceans?" calls out for comparison with Earth's neighbors, Mars and Venus. Both of those planets have lost their water, at least their surficial water. Although Mars may have had some tectonic activity in the past, it is increasingly clear that neither it nor Venus is tectonically active today.

Venus's geology is complex and still poorly understood, but we are nearly certain that it has no plates at all, or only a single plate, and that it lacks spreading centers and subduction zones analogous to those on Earth. So this "sister planet," in the familiar term, is geologically very different from Earth. Recent exploration of Mars has produced astonishing discoveries, one of which is that this now-dry planet once had considerable water, as shown by the fluvial erosion of impact craters and the layers of sedimentary rock found by the Mars Exploration Rovers. The Martian surface reveals scattered areas where groundwater has seeped out. However, the overall geomorphology of Mars— in particular its cratered Moon-like highlands—shows that this planet never had oceans comparable to those of Earth, which average about 3,000 meters in depth. And, like Venus, Mars never underwent plate tectonic activity. For a while the northern plains were considered to be the site of former ocean basins with sea-floor spreading, but that suggestion has been discarded in light of recent studies. The enormous number of high-resolution images of Mars from the Global Surveyor and Odyssey missions, and high-resolution laser altimetry, have shown that the northern plains have many subdued craters. However, no subduction zones or folded mountain belts, and apparently no plates, have been detected. As in the case of Venus, either there are no plates or there is only one.

To summarize: Our two "terrestrial" neighbor planets share certain geological similarities—no plate tectonics and no oceans. The Gaia concept now becomes relevant. It is clear that Earth's deep oceans heavily influence its plate tectonics. This influence is expressed in several ways.

Earth's Tectonic Plates

In plate tectonics, what moves the plates? Fundamentally, the energy source is heat from Earth's interior. Some of this heat is primordial, from the early Earth's formation by accretion. Much heat comes from radioactivity. A novel suggestion by Marvin Herndon (2010) is that Earth's core is a nuclear reactor comparable to the natural reactors found in the Precambrian uranium deposit in Oklo, Gabon.

One answer to the question of what drives the plates" is "ridge push," which refers to compression outward from the spreading centers, such as in the East Pacific Rise. This clearly is part of the answer, as shown by the widespread compressive stresses now known in large stable continental interiors. Gravitational sliding probably also contributes. The overall source of ridge push is uprising magma from the mantle, erupting as lavas at spreading centers. However, it is now known that ocean water is essential for the volcanic activity of spreading centers. Water lowers the temperature of initial melting of the underlying mantle. Oceanic water also promotes plate motion by increasing the cooling of the outgoing oceanic lithosphere. The circulation of ocean water through the crust is a process intrinsic to plate tectonic activity (Lowman 1976). Similar cooling contributes to "slab pull" in the subducted crust, promoting the rate of plate subduction. Other water-related factors contribute to plate movement, such as the conversion of basaltic crust to high-density metamorphic rocks such as eclogite. Hydrothermal circulation—movement of hot water—is important in all phases of oceanic plate motion, however, from start to finish.

Returning to Gaia: It is clear that plate tectonic phenomena would be enormously slowed, if not stopped completely, if Earth were suddenly to lose its oceans. Plate tectonic phenomena and oceanic activity ceased long ago on Venus and Mars, if indeed those planets ever experienced either. Margulis and Lovelock (1974) have, of course, answered the question "How has Earth kept its oceans?" They have asserted the role of an emergent planetary homeorhesis stemming from combined biotic and physical conditions, which have kept global temperatures in the range of liquid water, between 0 and 100 degrees Celsius, with the occasional exception of "snowball Earth" (or "slushball Earth") events (Hoffman and Schrag 2000). It is not likely that all of Earth's ocean water froze completely, even at the height of one of the pre-Phanerozoic ice ages, either the Marinoan (550 mya) or the Vangian (625 mya). Production of methane (CH_4), carbon dioxide (CO_2), and water had resulted in the now-familiar greenhouse effect. Other processes, such as rock weathering promoted by fungal and bacterial erosive processes, also have figured in regulating global temperatures.

As Euan Nisbet (1987) and Don Anderson (1992) suggested, a runaway greenhouse may have been prevented, in part, by the formation of limestone by marine organisms. That process, powered by photosynthesis, removed much of the CO_2 greenhouse gas from the atmosphere.

Without it, temperatures on Earth would probably have become high enough to force the oceans to "boil away," in Anderson's words.

Gaia is now a well-tested theory. The most developed example is the demonstration of the role of sulfur gases produced by *Phaeocystis* and other marine algae in the production of clouds over the open ocean, and their consequent temperature modifications—the so-called CLAW hypothesis (Charlson et al. 1987; see chapter 8). It appears that Earth's tectonic evolution has also been fundamentally affected by life for at least the last 2,000 million years. The reasoning behind this statement is discussed in the following section.

Duration of Plate Tectonics

When did Earth's plate tectonics begin? This question is still unanswered. Many geologists think the phenomena date back at least 4,000 million years, to the early Archean eon. However, they can be traced back to the early Proterozoic eon, about 2,500 million years ago, with much more certainty. The evidence derives from the now well-accepted theory, based on a wide range of geologic and geochemical data. Earth's geologic evolution underwent a major change in style at the Archean-Proterozoic boundary, about 2,500 million years ago. Most common geological structures and rock types associated with plate tectonics and increasing crustal stability appear in the record at about that time. For example, biogenic sedimentary rocks, such as limestones, marble, quartz, and sandstone, became abundant early in the Proterozoic Eon. Granites, clays, mudstones, and shales became more abundant. In summary, good geological evidence exists for Lovelock's assertion (1991) that surface conditions on Earth have, for most of geologic history, at least since the Proterozoic Eon, been modulated by life. The salient features of Earth subject to life's regulation (Lovelock 1988) are a mean planetary surface temperature that never exceeded the bounds of liquid water over the entire planet and a volume partial pressure of the highly reactive atmospheric oxygen gas of 20 ± 5 percent for at least 542 million years. The mild alkalinity of ocean water, the failure of all sodium chloride and other ocean salt to ever precipitate out in Earth's history, and the maintenance of detectable methane, sulfide gases, and hydrogen in Earth's atmosphere simultaneously present with oxygen are other clues to the long-term perpetuation of the Gaian regulatory phenomenon on the planetary scale. We suggest that this Gaian regulation extends to the biological modulation of Earth's tectonic plate activity too (Lowman 2002).

11
Evolutionary Illumination

Peter Warshall

Ancient life sensed light. Here Warshall traces life's interaction with light, from the earliest light-seeking bacteria and cells (whose entire bodies behaved as roving eyes) to animals seeking orange-colored vitamin-rich plants.

The Sun showers the solar system with electromagnetic energy, including cosmic rays and radio waves. One segment of the electromagnetic spectrum contains the signals for life's photosensory systems (Endler 1992). The photosensory segment of the electromagnetic spectrum extends beyond the visible range (that is, the range, between about 400 and 740 nanometers, that we can see in the right conditions). Various prokaryote strains and eukaryote species have evolved photoreceptors that utilize near-ultraviolet, near infrared, and polarized light, which humans cannot see without technology. I refer to the broader segment of the electromagnetic spectrum sensed by life as "biospheric light." It extends from about 200 nm to almost 1,100 nm, and includes polarized light.

A central task of the visual ecologist is to put the story of the planet's photo-environments together with life's photosensory systems. In large part, this photo-biological story explains how life can receive, evaluate, and act on only certain wavelengths and frequencies of the electromagnetic spectrum. It describes the photo-protective, photo-receptive, and photo-electro-biochemical history and "personalities" of cells, tissues, and organisms sensitive to biospheric light.

Over the roughly 3,600 million years of life's existence, certain photons have caused damage to the biomolecular structures and biochemical reactions required by life. Life evolved chemical sensitivity to these photons, including diverse modes of protection and escape from potential damage. If an organism was able to absorb photons without self-destruction or severe damage, the solar energy was free and of high quality. Sensory systems responded to light impingement by absorbing photonic energy and converting it to electrochemical energy. The electrochemical energy has been utilized to create life's biochemical energy and to discriminate contrasts, colors, movement, and distances. Minute differences in light qualities and quantities, in light ray directions, and in temporal cyclic patterns have stimulated and selected for evolutionary photosensory changes.

What are the commonalities of photosensory systems throughout life's kingdoms? How is biospheric light, along with a limited suite of biopigments and structural colors, involved in photobiological evolutionary innovation? How did "biospheric light" sensitivity change from the prebiotic molecules of the Hadean, photosynthetic prokaryote absorption of specific wavelengths in the Archean eon, and the refinement of spectral tuning in the Proterozoic to the explosion of reflective color perception by animal eyes in the Cambrian period of the lower Paleozoic (Cockell 2001)?

I would be the first to proclaim that a short essay on a 3,600-million-year history of light, life, and photosensitivity becomes outrageous because of the huge gaps in our knowledge. Historical reconstruction in general is plagued by methodological deficiencies. No scientific history of the solar spectral qualities that reached Earth through the ever-changing filters of various paleo-atmospheres has been written. Investigations of the photosensory diversity of protoctists have hardly begun. Little is known of how microscopic morphologies generate structural colors. My historical reconstruction here requires integrating information from planetary sciences, genomics, microbiology, photobiochemistry, cognitive sciences, and a dozen more disciplines. I am predominantly an animal ecologist. In addition, many of the planet's greatest human minds (e.g., Hsieh Hoh, Aristotle, Kant, Newton, Goethe, Wittgenstein) have attempted to understand the "affliction of color" (Goethe's phrase) and have left today's intellectual and scientific worlds dissatisfied. All I can say is that earlier understandings tended to focus on cognitive and anthropocentric concerns; few writers, if any, have attempted to overcome these intrinsic limitations and to reconstruct a photo-biological evolutionary history of the planet.

Basic Photosensitivity

All photosensitive systems appear to be based on counting photons (moles sec^{-1} m^{-2}), not measuring solar energy (watts/m^2). Sampling and counting photons appears to be advantageous to life forms, in part because, without deviation, one photon excites one molecule. Counting photons simplified the evolution of pigments "tuned" to those specific bandwidths that best excited an electron to jump to a higher orbital. Bandwidth (a small interval of wavelength) absorption by a photo-activated pigment stimulated the evolution of photo-behaviors in members of all kingdoms of life.

Photosensitive systems have specialized in four kinds of environmental information.

First, they have responded to the presence of harmful photons. All forms of life on Earth have responded by screening out harmful photons, neutralizing powerful toxic molecules created by radiation, and using light and chemistry to repair harmed molecules. When living creatures became mobile, specific taxa responded by photophobic and phototactic movement.

Second, photosensitive systems have absorbed photons and transduced them for biochemical energy (by electron transfer or photo-isomerization) within the cell or sets of cells. The need for photons has led to many types of light-sensitive tissues, and to growth and movement of many kinds of organisms toward biospheric light. The outstanding example of energy transduction has been photosynthesis (conversion of light to chemical energy) in prokaryotes, protoctists (algae), and plants.

Third, photosensitive systems have transformed biospheric light into electrochemical information that may alter behavior. Photosensitive systems have evolved ways to obtain spatial and temporal information about how to respond to local environments. These abilities exist in all kingdoms of life. For instance, some prokaryotes and some protoctists are photophobic, phototaxic, and photokinetic; all plants and algal protoctists display photo-morphogenetic and photosynthesis-enhancing behaviors. Image-forming vision has evolved in several animal lineages.

Fourth, certain taxa have been able to code responses to the periodic changes of solar radiance (and the reflected solar radiance of moonlight) into their genomes. They have internalized reactivity to photon counts and wavelengths as biological clocks, as discussed in chapter 9. Photo-period sensitivity exists in members of all kingdoms of life.

Pigments

For a painter, a pigment is a colored material ground fine (about 4 micrometers) and mixed into linseed oil, water, or another medium. The pigment absorbs and reflects select wavelengths of light. The light rays that are reflected rather than absorbed give the painted surface its color. In living creatures, a chromophore is a molecule that absorbs a narrow bandwidth of biospheric light. Chromophores are composed of organic carbon ring compounds with alternating double and single carbon-carbon bonds. When embedded within a protein, they are called pigments. In turn, the pigment is embedded in membrane, in a part of an organelle, or in some other component of a cell. As in painting, the wavelengths reflected will be perceived by us as the "color" of the organism.

In almost all photoreceptors, biospheric light must eventually contact a pigment if it is to change an organism's behavior. Bio-pigments are the interface between the outer environment and the inner world of an organism. In order for light to interact with a pigment, the pigment must be housed in a cell large enough to "collect" the incoming wavelengths. Wavelength amplitudes set a lower limit to the size of photosensory cells. If a wavelength is longer than the cell, some part of the wave will be outside, and the cell will not be able to absorb it effectively. For a wavelength of 500 nm (green), for instance, a practical size for a cell (or the cell plus inter-cellular space) to effectively sample photon flux is between 1,000 and 1,600 nm, about two to three times the amplitude of the wavelength.

The chemical diversity of pigments as the relevant units in the history of life and light is limited. Reflecting this ancient history, only three types of molecules make up the "palette" of bio-pigments. They are molecules specialized for light absorption. Terpenoids (carotenoids, certain quinones, melanin, retinal-related proteins synthesized from carotenes), tetrapyrroles (e.g., metal chelated components of hemoglobin, chlorophyll, and phytochrome), and phenolic compounds (flavonoids) are the major classes of chemical compounds in the history of life's pigments. (After the Cambrian, structural colors supplement pigment color. See table 11.1.)

In tandem, these three classes of molecules can now absorb over the complete biospheric light spectrum, from ultraviolet to infrared (table 11.1). Flavins, metalloporphyrins, and carotene absorb ultraviolet. These pigments, along with retinal-related compounds, chlorophyll, and other carotenoids, absorb blue. Bound carotenoids, flavins, and retinal-related rhodopsins absorb green and yellow wavelengths. Bacteriorhodopsin and a phycocyanin absorb yellow and orange.

Chlorophyll, phytochrome, and other hemes absorb red and far-red light. Certain bacterial chlorophylls can absorb infrared. Melanins can absorb all wavelengths, creating a black reflection, or partially absorb

Table 11.1
Biospheric light wavelengths.

Color	Approximate wavelength range (nm)	Representative wavelength	Representative energy (kJ mol^{-1})
Ultraviolet-C	200–280	254	471
Ultraviolet-B	280–315	300	
Ultraviolet-A	315–400	380	
Violet	400–425	410	292
Blue	425–490	460	260
Green	490–560	520	230
Yellow	560–585	570	210
Orange	585–640	620	193
Red	640–740	680	176
Infrared	>740	1400	85

in the shorter wavelengths, reflecting brownish orange or red. Given its biomolecular constraints, each of life's kingdoms has evolved toward becoming photosensitive to the complete spectrum of available biospheric light. The photosensitivity spectrum in animals, for instance, has been covered by retinal-related compounds, and that in plants by the flavonoid-to-phytochrome complex (figure 11.1).

Bio-pigments do not just display colors as in painting. Bio-pigments multi-task. A bio-pigment may selectively screen wavelengths that enter a cell; it also may strengthen tissue, act as an anti-oxidant or an antibiotic, modify a larger animal's heat balance, facilitate photon capture and optical resolution, or stimulate or govern temporal growth and development patterns. A bio-pigment may act as a component in a photosensitive system to detect and/or discriminate colors, edges, and forms, and may trigger emotional responses. Melanins, for instance, strengthen tissue, transduce gamma rays to biochemical energy in fungi, regulate heat absorption, darken a surface for camouflage, provide tonal contrast to white and red, or enhance iridescence. Carotenoids aid photon capture, transfer photons to chlorophyll, protect cells against oxidative harm, channel food webs, and act as a colorant for advertising toxicity or desire. (It is crucial to understand that the function of a pigment, or that of a structural color, may be a compromise among multiple tasks.)

In short, photosensory systems, depending on the taxon and the biospheric light niche, evolved to perform the following functions, at least:

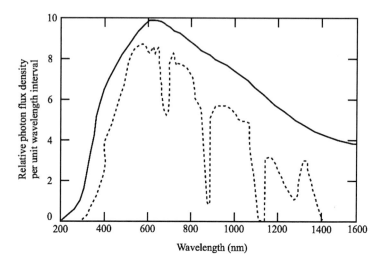

Figure 11.1
Biospheric light (broken line) in the atmosphere differs from solar radiance (solid line). In the range 200–400 nm, the difference is ozone filtering of UV. The dips between 700 and 1600 nm come predominantly from absorption of carbon dioxide and of water vapor. The solar radiance curve is theoretical and is based on an incandescent black body at 5700°K. The "typical" day is defined as clear skies at sea level.

• transduce photonic energy into chemical energy for motility, food production (photosynthesis), environmental information, and emotional response
• count photons (brightness)
• compare photon counts and polarization (directionality, temporality, dark/bright contrast)
• selectively absorb and reflect wavelengths to discriminate between by color, hue, or tonal contrast
• selectively absorb specific photons and transform them into heat as a protective or warming device.

Sophisticated, recently evolved photosensitive systems may use all these anatomical and physiological approaches in combination.

Sun-Earth Coupling

Biospheric light and photosensory systems start with the Sun's light. From life's point of view, the Sun is a single incandescent, long-term, reliable emitter of photons. But what if the Sun were an unstable star? Could life have evolved light-sensitive systems? How unstable can solar radiance become and still accommodate photosensory evolution? We do not know, but we can approach the question because solar constancy has been investigated by means of satellites.

The Sun would be less reliable if it fluctuated wildly in its output of frequencies, wavelengths, and/or density flux of photons. Large unpredictable variations in the composition of the solar spectrum due to solar flares, sunspots, and long-term cycles of the photosphere might have undermined the evolution of reliable photosensitive systems. A neutron-star merger or a supernova might have caused significant changes in the solar radiance reaching Earth's surface. More temporary but significant changes in the biosphere's filtering capacities probably have occurred after impacts by bolides (e.g. asteroids) and comets. Because investigations of bolide impacts have focused on infrared and other heat losses rather than on changes in overall spectral composition of the light reaching Earth's surface, we know little about their effects on photosensory systems. Recently measured spectral output, however, has been remarkably constant: the Sun's luminosity has not fluctuated wildly in the last 3,600 million years. The Sun's recently measured output has been remarkably constant. Satellite readings (recorded outside the atmospheric envelope) indicate that, in the human visible and infrared bandwidths, the Sun's annual photon flux densities have varied by less than 1 percent. (Specific ultraviolet bandwidths of the solar spectrum vary as much as 7 to 10 percent.) These variations in solar emission are small relative to the planet's night-day, seasonal, and latitudinal variation. This solar emission, with its small variation in intensity, is called the *solar constant*.

This is not to say that the Sun's output hasn't changed. Astrophysicists are sure that solar radiance has increased by at least 25 percent in the last 3,600 million years. Given the slow pace of change and the shielding properties of the atmosphere, photosensitive systems have been able to accommodate to this steady increase in photon flux. Astrophysicists do not know if the ancient Sun was simply smaller, with the same spectral composition, or if the Sun's spectral composition differed significantly. Most astrophysicists think that the solar spectrum reddened (that is, increased the photon flux density of wavelengths around 680 nm) as its radiance increased, but yellow-green wavelengths (about 490–595 nm) have apparently always predominated. The yellow-green predominance may have influenced the evolution of the earliest bio-pigments.

The largest variations of photon flux and spectral composition—the photon flux densities and wavelength differences between day, twilight, night, and dawn—are attributable to Earth's spinning. From midday through nighttime, light intensities can vary eight or nine orders of magnitude. At the time of life's origin, a complete day/night cycle took less than 14 hours. It took more than 2,500 million years to establish the 24-hour day. How shorter days affected early life has not been estimated. Despite the variation in daylight/ nightlight proportions and intensities, the long-term temporal precession of daily, seasonal, and annual photochanges gave rise to photosensitive endogenous biological clocks.

Photosensory systems that can cope with the magnitude of daily change in photon flux densities have not evolved. Instead, photo-systems have specialized—scotopic (monochromatic vision in low light) vs. photopic (vision, including color vision, under well-lit conditions). The cartoon version of this specialization contrasts butterflies by day and moths by night, or cones for bright light and colors and rods for low photon flux. Daily and seasonal Earth spin probably "pre-adapted" photosensory systems to cosmic changes from meteor and asteroid impacts.

Finally, the elliptical orbit of Earth around the Sun remains constant enough that the Sun is always at optical infinity: its rays reach us in parallel lines. The Sun-Earth distance has allowed the evolution of photoreceptors based on relatively simple optical geometry and, in the ocean's photic zone and on land, has made sunlight one of the best media for determining direction and distance, and for garnering information from the size and shape of shadows.

In short, three solar-system constancies have been of major importance to photosensory evolution: the constancy of the Sun-Earth orbital distance, the constancy of the Sun's spectral emissions, and the eventual constancy of Earth-spin time periods. Astrophysicists suggest that, over at least the last 542 million years and probably over the last 1,500 million years, Earth experienced limited solar spectral variation and no spectral catastrophes in bandwidths that affected life severely. Much work remains to be done on reconstructing paleo-atmospheres and their effect on the bandwidths available to evolving photo-receptive organisms.

The Niches of Biospheric Light

Biospheric light, not sunlight, travels to life's photoreceptors. Solar radiance becomes biospheric light by passing through multiple media that alter the intensity, the direction, and the quality of the light available. The media include all the layers of the atmosphere; the reflections, inter-reflections, multiple absorptions, and transmissions that solar rays encounter in the biosphere ("ambient" light); and the depth and quality of the planet's waters.

The transformation of solar photon flux and the sorting and altering of the Sun's wavelengths done by the atmosphere, by water, and by life itself create "niches" with different light qualities. The deep sea floor may be dark—that is, it may have no solar photon flux. Forest shade is yellow-green, woodland shade is darker (lower photon flux) with more blue and ultraviolet light, and cloudy weather whitens all habitats.

Atmospheric Niches

Today's atmosphere blanket of gases, aerosols, dust, and water vapor extends hundreds of kilometers into space. It is the first barrier and filter-changing solar radiance (figure 11.2). Today's atmosphere is relatively transparent to the wavelengths that stimulate photo-biological responses. About 99 percent of the Sun's energy long wave of 310 nm (ranging from UV-B to near infrared) penetrates to the troposphere. In the present-day atmosphere, ozone reduces ultraviolet (200–400 nm) from 5 percent to 2 percent of the total solar flux. Human visible light

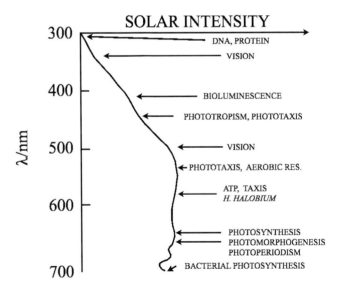

Figure 11.2
The photo-biological spectrum, with a solid line representing the solar spectrum and the arrows indicating photo-responses at the wavelength maximum of photosensitivity. The actual spread of photosensitivity for each pigment is not shown. To the right of the solar spectrum diagram are light-emitting processes.

(400–700 nm) increases from 25 percent to 45 percent (and can be higher). Infrared light decreases from 63 percent to 53 percent (and can be lower). Carbon dioxide and water vapor filter out specific wavelengths of red and infrared. This remnant spectrum of solar radiation comprises the biospheric light of photosensitive organisms (Lean 1997).

Biospheric light contains another signal, too often ignored because humans cannot see it. The skylight and cloudlight of Earth are polarized. Underwater and surface reflected light may also be polarized. (Direct sunlight is weakly or not polarized.) Near-UV is frequently the band-width used to detect polarized light because UV is a short, high-energy wavelength that penetrates the atmosphere more consistently than long wavelengths. (In addition, long wavelengths, greatly affected by clouds and moisture, vary ten times as much as UV in a day.)

Short wavelengths are maximally polarized and long wavelengths weakly polarized such that those species that can detect and respond to polarized light usually see beyond the visible spectrum and into the ultra-violet. The sky thus provides a photo-signal, due to polarized light, that has well-documented effects on protoctists and on animals. Dragonflies, for instance, use polarized light to stabilize flight, water striders to reduce glare, birds for celestial navigation, bees for local navigation, and under-water crustaceans and certain insects for increased object resolution.

Water Niches

Since photosensory systems began in the seas, the Hadean ocean was our first light niche and set the basic parameters for the cellular and tissue matrices of photoreception. Water has specific physical properties (Denny 1993). Water absorbs and scatters incoming light more quickly than air. Water diminishes photon flux densities and narrows biospheric light's spectral range. Two hundred meters below the sea's surface, all light is absorbed, even by clean water. Of course, in occluded water the photic zone may be only a few millimeters. Photosensitive, solar-driven pig-ments in prokaryotes, protoctists, plants, and animals are useless below this photic zone. Since the average depth of the contemporary ocean is 3,000 meters, such strong attenuation selects for bioluminescence, echo-location, or vertical migration in 95 percent of the ocean's volume.

In today's water, "seeing" becomes difficult over long distances. Dis-solved organic matter, chlorophyll, plankton, and suspended solids in water diminish photon flux densities and narrow bandwidths. At dis-tances longer than about 30 meters. Most photosensory systems are unable to detect contrast and color in water. Natural selection has favored sound-based sensory systems over photosystems.

Photon penetration into water is best between 300 and 550 nm (UV-A, violet, blue, and green wavelengths to human eyes). Blue light penetrates the deepest. About 90 percent of incident blue light (460 nm) penetrates to 30 m, about 78 percent to 50 m, and none to 200 nm. Within this band of wavelengths, the evolution of the earliest active pigments occurred. All major photosynthetic chlorophyll pigments absorb within the blue segment, the segment least attenuated by pure water. On the other hand, UV (326 nm) flux densities, even though more energetic, attenuate by about half at 30 meters and to 25 percent of their original flux at 50 meters.

Red wavelengths fare worst in water. For wavelengths of 700 nm, there is no photon flux at 15 m and very little at 8 m. As scuba divers can verify, a red air tank will quickly appear gray as the diver wearing it descends. Below the "red light zone," the only creatures with photosensitivity to red wavelengths are those that produce red light (by bioluminescence) for predator-prey or courtship purposes. Pigments (such as chlorophyll) that also require red light must have evolved near the water's surface and remain restricted to that zone.

In summary: The bandwidths affecting life are not the same as the spectrum of the Sun that surrounds the planet outside the biosphere. Light passes through and is modified by transparent and translucent materials—atmospheres, oceans, leaves, puddles, or butterfly wings. Light changes each time it is reflected from the landscape or from the body surfaces of living organisms. Each attenuated photon flux, with its unique spectral qualities, defines a photo-niche. Photo-receptors for each photo-niche evolved and complexified (especially during the Paleozoic era), with specialized filters, reflectors, amplifiers, and wave guides that occur right before the photo-receptive pigment.

The next section sketches the major environmental and biological events in the photo-biological history of the planet: prebiotic molecules, Archean bacterial photosynthesis, bacterial chlorophyll, protist symbioses and accessory pigments, invasion of land by embryophytes, and the revolution in image formation that occurred during the Cambrian period.

The Hadean (before 3,900 mya)

Prebiotic Molecules

Unless life was seeded from another planet, the precursors to the first photosensory cells were shaped by Hadean solar radiance. At the high-energy, short-wavelength end of the spectrum, nucleic acids (RNA,

DNA) and aromatic amino acids readily absorb UV-B (280–315 nm). UV's high energy causes lesions and dimers, destroying molecular structure, and can create highly reactive radicals, especially in water.

At the other end of the spectrum, infrared radiation (and other sources of heat) causes rotational and vibrational movements in molecules that prevent them, especially enzymes, from effectively participating in life's biochemistry. Near-UV and near-infrared define the safe interval of electromagnetic frequencies and wavelengths of biospheric light. UV and IR radiation must have destroyed many experiments in the prebiotic generation of organic molecules. (See chapter 2.)

The prebiotic danger from ultraviolet (less so for infrared) has been a topic of lively discussion because of ozone pollution. On one hand, some speculate, on the basis of stars similar to our Sun in age and mass, that the Archean Earth could have experienced 10,000 times as much UV irradiance as the present-day Earth experiences. On the other hand, our Sun may not have been typical. An early-Earth sun with 25 percent lower solar luminosity may have also reduced UV irradiation by 25–35 percent. In addition, a paleo-atmosphere with a much higher percentage of carbon dioxide probably absorbed all UV below 200 nm. Volcanic gases and aerosols, methane smogs, sulfur-dioxide or hydrogen sulfide hazes, or even an aldehyde haze—all possible in paleo-atmospheres— would have reduced UV significantly. In the intertidal waters, sulfur, iron, salt, and sediments such as nitrate in solution could have been additional physical UV screens. Reasonable estimates now project two to three orders of magnitude more UV-B at the time of life's origin. This translates into a required protective screen of 30 meters of clear ocean to match the protective level of today's atmosphere.

In short, the reducing atmosphere of the prebiotic Earth had little oxygen, intense ultraviolet-B, significant amounts of ammonia, methane, and CO_2, spotty but available organic molecules, and wavelengths of solar radiation that, if harvested, could be used to power prebiotic synthesis, CO_2 reduction, and fermentation. Among the organic molecules available to early live were three likely candidates for the first photoreceptor pigment: flavonoids, porphyrins, and retinal-based compounds.

The Archean (3,900 mya)

Wavelengths absorbed, not reflected, played the most important role in the evolution of early photosensitivity. Pigments absorbed ultraviolet and other harmful radiation, and absorbed and transduced useful photons into biochemical energy. Flavonoids are pigments found in all cells. In archaebacteria, they are not fixed in location to an organelle or a cell membrane. They absorb UV and blue wavelengths, and release some of harvested UV energy as fluorescence and heat. Some flavins attach to the enzyme photolyase, capture blue light (425–490 nm), and transfer the acquired energy to the enzyme molecule. (Very few pigments can be found embedded in enzymes.) Photolyase, in turn, repairs UV-damaged nucleic acids. In these flavins, we see the ancient movement of simple photo-screening to bandwidth tuning, and a more complex photochemistry.

Retinal-based pigments have been found in the halobacteria presumed to be representative of the Archean. These prokaryotes contain a membrane protein (called an opsin) to which is attached a (retinal) prosthetic group from one of vitamin A's aldehydes. Together, they form a variety of the pigments called rhodopsins (figure 11.3). Rhodopsin transduces light in the photosynthesis of the energy molecule ATP in halobacteria, in which both sensory and energy-transduction rhodopsin can be found in the same cell. Halobacterial photosynthesis (photophosphorylation of ATP by use of rhodopsin) evolves independent of the far more familiar chlorophyll-based photosynthesis. Much later, after halobacteria evolved rhodopsin, pigmented proteins appeared in the retinas of vertebrate eyes. Rhodopsin-based photosynthesis that generates energy but does not produce food differs entirely from cyanobacteria-plant photosynthesis.

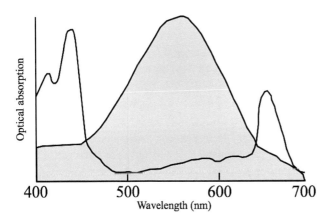

Figure 11.3
The absorption spectrum of bacteriorhodopsin (shaded) may have set the environmental circumstances for the evolution of bacteria that use chlorophyll-carotenoid pigments in light absorption. Chlorophyll *a* (black line) avoids (reflects) yellow-green (500–600 nm) and absorbs the red and blue regions of the visible spectrum on either side of it.

The pigment absorbs yellow-green (about 570 nm) perhaps where, at least at present, it is the strongest light emitted by the Sun that reaches Earth. Some evolutionary historians believe that the Hadean oceans had enough turbidity to absorb the blue wavelengths. Others speculate that the blue was already harvested by the flavins for photoreactivation of nucleic acids.

Such presumably ancient bacterial rhodopsin pigments absorb light very efficiently and remain photo-stable (they are not destroyed when they absorb light); they are also thermo-stable. Some are photo-labile, can isomerize, and can transduce solar energy into electrochemical energy. Archaebacteria with rhodopsins do not seek light, but some retreat. They avoid UV by swimming away. Both their sensory rhodopsins and their energy-transducing rhodopsins influenced photosensory evolution.

One sensory rhodopsin found in a *Halobacterium* sp., for instance, isomerizes in a manner similar to the rhodopsin-based visual pigments in animals. If a direct line is traced from this rhodopsin-bearing archaebacterium, the molecular basis for animal vision has evolutionary roots from more than 3,000 million years ago.

Chlorophyll is a cyclic tetrapyrrole that may have begun prebiotically as a metal-free porphyrin. The chemical class romantically called the "pigments of life" is widely distributed in most life forms. It includes such other cyclic tetrapyrroles as cobalt-chelated vitamin B_{12}, iron-chelated hemoglobin, and cytochrome. Anaerobic chlorophyll-based photosynthesizers probably originated close to the water surface by protecting themselves from UV in sediments. Chlorophyll, a manganese-chelated tetrapyrrole, absorbs red and blue and fixes carbon dioxide in the manufacture of sugars.

The molecular underpinnings of nearly all pigment color and photosensory systems had evolved by 3,000 million years ago, long before plants. The tensions between photosensory tasks had begun. Cells needed to selectively filter some wavelengths of light for survival, photoreactivation of DNA, and photosynthesis of ATP, and to produce energy. But other wavelengths that might cause cell damage or cause too much interference with the photo-signal had to be simultaneously excluded. "Spectral tuning" by the pigment-protein complex improved photoreception before the evolution of any eukaryotes. In the Proterozoic eon, symbiogenetic acquisition of photosynthesis as permanently undigested food by motile, sneaky, persistent protoctists filled the waters of the world, from the tropics to the poles.

The Proterozoic (2,500 mya–542 mya)

By 2,000 million years ago, an oxygenic form of photosynthesis that required chlorophyll and carotene had evolved in photosynthetic bacteria. Carotene acted to quench excited states of oxygen. The oxygenic bacteria excreted oxygen and increased its concentration in the sediments and the oceans. Their proliferation threatened most other life forms. The chlorophyll-based oxygenic bacteria began to change the world forever. Since the Proterozoic eon, Earth has experienced inexorable oxygenation and greening.

The retreat of the purple photosynthetic bacteria left the middle of the spectrum open. The green absorption gap facilitated the evolution and proliferation of many new autotrophic prokaryotes, later protists with "accessory" pigments. Accessory pigments did not transduce photons to biochemical energy themselves. Accessory pigments transferred electrons from the mid-spectrum to oxygenic chlorophyll. Accessory pigments included new variants of chlorophyll and tetrapyrrole pigments, always accompanied by protein-bound orange and yellow carotenoids.

Competition for unused or less used bandwidths, like today's competition for radio bandwidths, led to filling up of the visible electromagnetic spectrum. Cyanobacteria absorb a larger portion of the spectrum than green or red bacteria. Red algae originated by symbiotic fusion of heterotrophic protists with cyanobacteria. Rhodophyta (red algae), in turn, absorb more than cyanobacteria. In fact, red algae absorb almost all wavelengths of biospheric light and, despite their name, reflect brown and burnt sienna.

During the Proterozoic eon, the expansion of photoreceptors occurred, largely as a consequence of endosymbiosis among protoctists. "Eye spots" or stigmas that improved directional reactions to specific wavelengths of light evolved in many phyla (e.g., dinomastigotes and euglenids). Motile protists transduced light into chemical information that distinguished brightness from darkness, UV-harmful zones from UV-safe zones, infrared-intense niches from IR-safe niches, and so on.

The protoctists generated complex photosensory systems. The single-cell *Erythropsidium pavilardii* has a complex ocellus, a prey-detecting "eye" with a lens and a fluid-filled chamber underlain by a light-sensitive pigment cup. The pigment cup is a modified plastid and transduces harvested photon-derived energy directly to the undulipodia. The lens changes shape. The pigment cup moves freely about the lens. The whole ocellus protrudes to better detect prey.

The Phanerozoic (542 mya to Present)

Whereas the Hadean told a story of absorbed and harvested photons, the Phanerozoic tells a tale of reflected wavelengths. The Phanerozoic could be called the Age of Color Patterns. The land greened, and the colors of coral reefs intensified. The terrestrial canopy set the stage for pigments to exploit the far-red and for consciousness to deal with shadows. A new technique for creating reflected colors called structural colors spread worldwide. Reflected color reorganized food webs, both to broadcast and to camouflage the moving bodies.

During the Phanerozoic, animal, plant, and fungal species evolved color patterns and behaviors that were extensively reliant on discriminating vision. Sensitivity to hues, shades, and shadows fostered proficiencies in awareness and memory (e.g., "color constancy") as well as in discrimination and interpretation of tonal contrasts at every stage of life. Recognition of eggs, young, and adults, courtship, mating, recognition of sources of fruits, pollen, and nectar, camouflage, protection, and ambush all involved visual sensitivities. The photosensory systems of animals that interpret the tricks of reflected light evolved in tandem with neuronal wiring and the brain.

Greening of the Air beyond the Sea

At the beginning of the Silurian period, 439 million years ago, green algal relatives in association with fungal and prokaryotic symbionts moved out of sea water and fresh water, onto the land, and into the atmosphere. They took advantage of increased photon flux densities, more easily diffusible gases, and less competition for space. Ultraviolet radiation was no longer a nearly insurmountable problem, because stratospheric ozone now blocked more UV-B. By then, life had successfully responded to ultraviolet for more than 2,000 million years. (The early plant and fungal pioneers required strategies for nutrient intake and storage, water retention, and structural support in the far less buoyant medium of air.)

During the Devonian period (408–360 mya), embryophytes with roots, stems, and leaves evolved. As the land greened, green beings grew vertically toward the Sun and away from hungry ground-dwelling animals. By the end of the Devonian, forest canopies flourished and, once again, life shaded life.

Chlorophyll-based ferns and vascular plants removed blue and red in the canopy. An ancient algal pigment that blossomed in the Devonian was able to compare red light (which indicates an open canopy) against far-red light (which is reflected by leaves and indicates a closed canopy). By comparing red against far-red, plants accelerated stem growth and, using chemical signals, brought their leaves into more favorable light conditions. The biological clocks of interacting plants synchronized to maximize photosynthesis. A proteinaceous pigment called phytochrome even influenced the orientation of chloroplasts inside plant cells to maximize light absorption. Phytochrome is purely photosensory. It does not harvest photons directly, as chlorophyll does.

A relatively unoccupied portion of the spectrum (the far-red, at 730 nm) spurred a new chapter in photosensitivity. Plant life employed a dual-control photosystem in which two distinct wavelengths stimulated the same internal physiological system in metabolism that sensitized plants to changing terrestrial light conditions. Phytochrome pigments became major managers of plant color perception. Perhaps the subtlest component of vascular plant photosensory systems was concerned with maximizing reproductive output. By directing growth, phytochrome that did not change coloring enabled plants to harvest more photons. Phytochrome concentrations are so low that they are easily masked by other pigments. Like steroid hormones of animals, they induce dramatic changes in grown and behavior at minute concentrations.

Structural Color

During the lower Cambrian period, just before 515 million years ago, an evolutionary novelty that created reflected colors appeared in animals. The earliest fossil evidence of *structural colors* has been found in fossil trilobites and polychaete worms. Incident light was selectively reflected by thin films, diffraction gratings, and bodily protrusions and shapes. Structural colors, in contrast to pigment-reflected colors, may appear metallic and often show specular metallic properties. Some rapidly change hue (iridesce). One-, two-, or three-dimensional micro-structural arrays constructed from organic molecules (e.g., guanine scales in fish, chitin in insects, keratin in birds, collagen in mammals) are sources of structural coloration. Animal structural color generation has added ultra-violet colors.

Today, structural colors, sometimes combined with pigments, increase reflection (fish scales, tapeta), coloration, glitter, and iridescence (beetle elytra, bird feathers, butterfly scales, eye colors in tabanid flies, surface colors in coral-reef fish, skin in cephalopods and primates).

Animals and Eyes

The animal kingdom, more than any other, evolved elaborate photosensitive visual systems that increased growth and reproductive potential by improving the efficiency of courtship, mating, and brood care.

The evolution of the animal eye-brain relationship from pinhole to camera eye and beyond cannot be explained briefly. Eyes serve as wave guides, wavelength filters, and enhancers, as well as optical distance adjustors. They integrate information across multiple wavelength channels, use temporal cues to un-mask background noise, and exploit the directionality of peripheral photon flux to tease signals from noise. During the Archean and Proterozoic eons, photoreceptor systems harvested a wavelength that was identical to pigment's maximal absorption capacity. In the Phanerozoic, many pigments maximally absorbed wavelengths to the "side" of the incoming color with the most photon flux. In many circumstances, this change in the function of spectral tuning increases tonal contrast perception. It is indicative of the move from simple absorption of select wavelengths to a complex assessment of biospheric light. (See box 11.1.)

The Reorganized Food Web

The Phanerozoic witnessed a dramatic re-organization of food webs in photic zones. Unable to synthesize the vitamins they need for vision and survival, animals must obtain them by selective herbivory and/or carnivory. Also during the Phanerozoic, animals evolved powers to process colored images. In terrestrial communities and in shallow aquatic communities, pigment and structural color patterns became a "color commons." Except in caves, on dark nights, and in ocean depths, color became a signal that announced a living presence.

Animals cannot synthesize vitamins K, B_1, B_2, B_{12}, and A by themselves. These vitamins must be synthesized by prokaryotes, protists, or plants. Vitamin A, derived from carotenoids, is the molecular basis for retinoid, which is critical to animal vision. Herbivorous and carnivorous animals must obtain carotenoids or vitamin A from their food. The polar bear, a classic example, hunts seals by sight. Seals eat fish, which eat smaller fish, which eat zooplankton, which eat algae or bacteria, which

Box 11.1
What is color photoreception and vision?

"Color" creates a photosensory pattern in a slice of time. Physically, the pattern begins as the distribution of the number of photons at each wavelength striking a given area (cell, retina, pigment) per second. This is a necessary, not a sufficient condition. If an organism can detect only one cluster of wavelengths, then it is hard to say that it sees color. Essentially, its world is monochrome. Even if a pigment absorbs at two wavelength maxima (as in chlorophyll), these two signals must somehow inform each other and change the behavior of the cell. Some biologists are comfortable saying that an organism is photosensitive to color only when there are two separate pigments or when there is a pigment-protein complex that can isomerize and respond to two different wavelengths (e.g., phytochrome). Color, in their interpretation, occurs only when the cell must sort out two different signals and compare their ratio. Some zoologists argue that there is no color vision until there is a neuronal mechanism to compare the photon fluxes and wavelength clusters from rods (or between rods) and cones. This definition is too phylogenetically limited. What properties constitute an "image" in photosensory systems has not been resolved.

synthesize carotenoids. Without dietary carotenoid, the bear would not synthesize retinoid and would not be able to hunt.

The need for vitamin A for vision is a major driving force in animal survival. Predators and herbivores search out food items with high carotenoids. Since carotenoids reflect red-orange-yellow (ROY). "Pigment harvesters" that became more efficient at spotting these colors had a survival advantage. Prey or herbage that could hide carotenoids (or could mix them with poisons) escaped predators and consumers.

Once the telltale ROY signal was "let loose" in the Phanerozoic environments, ripple effects occurred. Certain ripe fruits advertised carotenoids with ROY skins and attracted birds or mammals to eat them and disperse their seeds. In the Cenozoic era, songbirds utilized carotenoids not only for vision but also to color their plumage. Saturation of a male's ROY plumage signals a female that the male either has a good physiology for processing carotenoids or has a good home range with abundant sources of the pigment. The ROY colors jumped from announcing the presence of pigment to symbolic advertisement of flowers for pollination and fruits for nutrition.

With the advent of flowering plants during the Cretaceous, about 145–120 mya, angiosperm flavonoids were added to the ancient carotenoid pigment repertoire. Flavonoids signaled pollinator insects, and later birds and mammals, by providing colored and conspicuous beacons, landing strips, and nectar guides. Colors even changed within a single plant,: one color before and another after pollination; one for unripe fruit and another for ripe fruit; one for day and another for night.

Plant flavonoids first selected for in the evolution of lignin as wood for vertical extension of the biosphere, now returned to their earlier role as small molecule pigments. They colored cells in some flowers, fruit skins, bracts, leaves, and stems. Flavonoid colors evolved from simple creams and yellows to shades of white and advanced pinks, reds, and purples. The most complex colors appeared in ornate flowers that attracted animals.

The association of a particular bandwidth of reflective color with a desirable resource led to predators and herbivores with improved spectral tuning that recognized ROY organisms more easily. The ability to see a color generated a capacity to react to an object in shade or under a canopy as if its colors were in full sunlight. We humans do not normally view an apple in a shadow as purplish gray. If asked, we will say that the apple is red. Holding constant the "memory" or image of the sunlit apple's color, despite the myriad changes of biospheric light, has obvious advantages. "Color constancy," a term for the retention of a color in varied lighting, has been demonstrated even in moths.

As the carotene-based ROY colors became conspicuous, some animals avoided the use of precious carotene by substituting other pigments, flavonoids, red melanins, and structural colors. They hid carotene-based ROY colors in darker pigments, such as black and brown melanins, or confined their appearance to less conspicuous body parts, such as the stomach. Some added poisons so that the meaning of ROY colors now signaled "unattractive." Birds with toxic flesh advertise their poison with the same ROY colors that others use in courtship. Ladybug beetles, coral snakes, and Monarch butterflies use ROY colors as warnings that they contain poisons. Such double and multiple entendres increased the selection advantage of learning and education.

Aesthetic Zeal

Jerome Lettvin liked to say that natural selection created an "aesthetic zeal" among carnivores and herbivores to discriminate prey or plant resources from background lighting as well as to isolate certain color patterns associated with successful mating or with pollination. (Lettvin uses the word "aesthetic" in its Kantian sense, referring to a branch of metaphysics concerned with the laws of perception; see Lettvin 1973.) Aesthetic zeal selected for a discriminating awareness of significance among reflected colors. Is it attractive? Is it a warning? Does it signify anything at all? In most animals, this is a learned discrimination.

The coevolution of color discrimination and tonal contrast accelerated during the Phanerozoic era. Visual acuity, ability to process received bandwidths of color, the vagaries of biospheric light, and the survival of both the perceiver and the perceived complexified. Sophisticated visual systems to distinguish background from object and parts of an object from the whole object evolved. Some insects and vertebrates evolved green or brown integuments to match leaves (background resemblance camouflage) in order to foil the perceiver. Some evolved deceptive resemblance (mimics). Some continuously changed their appearance to match their background—for example, there are cephalopods that can generate 30–50 different surface patterns for rapid protective coloration. Some animals evolved body colors that disrupted their body outline or their three-dimensional conspicuousness to break up their self-image, and some counter-shaded their body to conceal telltale shadows.

In Earth's photobiological history, a small number of bio-pigments restricted the phylogenetic possibilities of photosensory systems (Cockell 2001). This biomolecular constraint selected for molecular diversity within each class of pigments. In photic communities, the scarcity of pigments encouraged symbioses, specific food web connections, and other social and community interaction. Structural colors expanded the "palette" of informational, reflected color based on new genetic foundations.

Living beings share the same biospheric light niches. They have cross-mapped their responses to reflective and transmitted color patterns and signals that are shared between species. Color-signal overlap between species helped integrate ecosystems and, in turn, fostered progressively subtler color patterns, visual apparatuses, discriminating awareness, and analytic evaluations of what is illusion and what is real. The complex ancient photosensitive systems that produce and respond to visible signals became an essential precursor to the human mind.

IV

Chimeras

12

Symbiogenesis in Russia

Victor Fet

Bacteria, the first individuals, sensed their environments, found sources of energy and matter to maintain and expand their growth, and synchronized with Earth's great biogeochemical cycles. As they did so, some of them merged to form integrated many-cell communities that became a new kind of cell. Here Fet unfurls the historical contributions of Boris Kozo-Polyansky and Konstantin Mereschkovsky to the great idea that cells with nuclei (such as our cells) come from multiple lines of bacteria.

Russian scientists pioneered the now-accepted notion that free-swimming amoebae and the cells of animals and plants derived from permanent symbiosis among separate forms—partnerships that had major consequences for evolutionary history. The word "symbiogenesis"—meaning origin (genesis) through living together (symbio)—was coined in Russia by Konstantin Sergeyevich Mereschkovsky (figure 12.1) in the early twentieth century. That co-evolved symbionts are crucial to biotic innovations in the evolutionary process is taken for granted in certain specialized fields of Russian biology. Although the early symbiogeneticists (Mereschkovsky, Famintsyn, and others) held an evolutionary view of the living world, all except Boris Mikhailovich Kozo-Polyansky (figure 12.2) ignored or denied Charles Darwin's (1859) mechanism of natural selection as the main source of new species, new organs, new tissues, and other novelties. Kozo-Polyansky's (1924) integration of Darwinian and other anglophone evolutionary thought with the work of his Russian "symbiogeneticist" predecessors remained virtually unknown in the West before 2000. Ivan E. Wallin, like all the other biologists who wrote in English about living associations of organisms of different ancestries, died ignorant of the magnitude and subtlety of the Russian technical literature on symbiosis as a source of evolutionary innovation. Wallin's

Figure 12.1
K. S. Mereschkovsky.

prescient 1927 book *Symbionticism and the Origins of Species*, with its invented term equivalent to the Russian idea of "symbiogenesis," remains little known and grossly underappreciated. My goal here is to bring to modern readers an appreciation of the early evolutionists who developed a vigorous biological literature based on the ideas of symbiogenesis (symbionticism). The concept of species, tissues, cells, and other levels of biological organization as chimeric carries with it immense implications for experimental science and even medical research. (See chapter 17.) Symbiogenetic views have been increasingly shown to be correct by modern methods in genetics, molecular biology, ultrastructural analysis (electron microscopy), and other fields.

Figure 12.2
B. M. Kozo-Polyansky.

The word "symbiogenesis" derives from Greek but was coined in Russia. It was introduced by Konstantin Sergeyevich Mereschkovsky (1855–1921), a troubled and biased man but a brilliant and prescient scientist (Sapp et al. 2002). The idea of symbiogenesis as a driving force in evolution was put forward by Mereschkovsky in his 1909 treatise *The Theory of Two Plasms as the Basis of Symbiogenesis, a New Study of the Origins of Organisms*, in which the famous Kazan University botanist clearly stated that association of genetically disparate and distant organisms provided the major mode of evolutionary change. Mereschkovsky's work was known in the West. Several of his important papers were written in German. His final publication, which was in French, attracted little attention among mainstream biologists (Sapp et al. 2002). His suggestions were considered far-fetched and unconventional; it did not help that he considered symbiogenesis as a pacifying and unifying force opposed to Darwin's natural selection, then customarily seen as a fierce and bloody struggle (Khakhina 1992).

Mereschkovsky's younger compatriot Kozo-Polyansky (1890–1957) published only in Russian. His 1924 book *The New Principle of Biology: An Essay of the Theory of Symbiogenesis* was never translated into any other language. Kozo-Polyansky—a botanist known for his works on plant systematics—spent most of his life in the Russian provincial city of Voronezh. His book was unknown to researchers of symbiogenesis until recently, when Liya Nikolaevna Khakhina briefly reviewed it in her detailed 1992 book on Russian botanists, including those sympathetic to research in symbiosis. *The New Principle of Biology* remained obscure even in Russia, where, in the ideologically stultifying atmosphere that prevailed from the 1920s to the 1950s, Kozo-Polyansky's maverick ideas were shrugged off as a non-scientific fantasy—an attitude quite familiar to symbiogeneticists elsewhere.

Kozo-Polyansky was aware of Paul Portier, Ivan Wallin, and many other contemporaries who studied symbiosis in an evolutionary context. Unlike them, Kozo-Polyansky was not an experimentalist. But as he reviewed the work of others, he made sweeping theoretical interpretations.

First, he connected symbiogenetic ideas to the evolution of life but, unlike his predecessors, pledged allegiance to Darwin's concept of natural selection (Khakhina 1992). He saw Darwinian "evolution via selection" in the evolutionary success of cells, associations, consortia, organs, and organisms that had evolved by symbiogenesis. (He did not draw strict lines between those levels.) For Kozo-Polyansky, a symbiogenetic cell, organ, or organism was more advanced and more adapted, and thus better able to survive. He wrote about selection among various symbionts within a consortium, which he called, as we now would call it, "evolution of symbiotic systems."

Second, Kozo-Polyansky connected symbiogenesis directly to Darwin's pangenesis, and to Darwin's famous statement (1883) that a living cell is a microcosm. He interpreted Darwin's pangenesis theory (gemmules located in all cells and organs) as a direct precursor of symbiogenesis. This resembled De Vries's "intracellular pangenesis," where nuclear genes ("pangenes") were in fact Darwin's gemmules but present only in the cell's nucleus. For Kozo-Polyansky, they were present everywhere and were inherited via infection or with egg cytoplasm. Today, this vision is brilliantly vindicated by our knowledge of molecular mechanisms of heredity that include chromosomes as well as viruses, plasmids, transposons, and other phenomena forming a continuum of vertical, horizontal, and symbiogenetic inheritance within and between genomes and species.

Third, and most unexpected, the corollary of Kozo-Polyansky's New Principle is the fundamental discontinuity of life known today as prokaryote-eukaryote dichotomy. He did not, of course, use these words, which are traceable to a 1925 paper by Edouard Chatton (Sapp 2005). But unlike Chatton, he correctly recognized the importance of symbiogenetic origin of the eukaryotic cell from prokaryotic ancestors (Margulis 1993). Ernst Haeckel, in 1904, had already included cyanobacteria among his Monera with the bacteria and suggested that plant cells evolved by symbiosis with blue-green algae (Sapp 1994). That idea had been developed further by Mereschkovsky, who had first suggested evolution of chloroplasts from cyanobacteria. Kozo-Polyansky logically completed this chain of thought: for him, a "true cell" (what today we call a nucleated or eukaryotic cell) was a system, assembled from heterogeneous parts. Thus, for Kozo-Polyansky, not only were mitochondria and chloroplasts (which he treated almost as equivalent to the zoochlorellae and zooxanthellae of hydroids and corals) symbionts; all other organelles in the cell were suspect—the centrosome, the blepharoplast, the Golgi apparatus, and even the nucleus. Moreover, since "cytodes" (Haeckel's term for bacteria and cyanobacteria) had no nuclei, they must, Kozo-Polyansky argued, be ancestral to all symbiogenetic combinations; thus, for him and his contemporaries, bacteria were living units but couldn't even be called real cells. Like other biologists in the early twentieth century, Kozo-Polyansky simply and paradoxically limited the definition of a cell to what we now call eukaryotic cells. Thus, he was the first to attribute this most fundamental discontinuity of life on Earth to the symbiogenetic, synthetic, nested-doll-style structure of the eukaryotic cell, and to its evolution from simpler, prokaryotic symbionts. *The New Principle of Biology* remains an important and remarkable publication.

13

From Movement to Sensation

John L. Hall and Lynn Margulis

Spirochetes attach with great tenacity to other cells, and may propel them 100 times as fast as they would otherwise go. Some spirochetes still move from the periphery to the inside of their partners. It appears to us that the genomes acquired in spirochete partnerships later evolved for sensing light and chemicals and perhaps even for moving thoughts around brains. Here is the news: Genetic and molecular biological evidence supports the notion that the oldest of the organisms that merged in the formation of the modern nucleated (eukaryotic) cell, and arguably the most important for intracellular motility, was, and still is, the remnant spirochete.

A huge difference between prokaryotes (eubacteria and archaebacteria) and other organisms (plants, fungi, animals) is "intracellular transport," the transportation system that animates the insides of cells. Vigorous sudden or slow movement, universal in eukaryotes, is entirely absent from bacterial cells, even large ones. Cell movement in prokaryotes appears to occur below a light microscope's limit of visibility. Cell eating (phagocytosis), drinking (pinocytosis), and reproduction (including the kind that reduces the chromosome numbers in the formation of sperm and eggs) all involve intracellular movement. And most eukaryotic cell movement requires microtubules and enzymes.

In plants, microtubule-based movements influence how the new cellulosic cell walls develop or how some plant cells elongate to many times their width. The sperm's swimming, the beating movement of the cilia that direct the flow of air in the trachea and lungs, and the epithelium's guiding eggs toward fertilization in the oviduct all involve movement associated with microtubules. Eukaryotic cell motility requires interactions of many proteins (microtubules, fibrils, membranes, molecular motors) that make up the transportation system appropriately called the cytoskeleton. The cytoskeleton always resides inside the nucleated cell.

Even though elements of it (axopods, recurrent undulipodia, axostyles) often form protrusions, the microtubule-based cytoskeletal system lies beneath the cell membrane (Chapman et al. 2000). No such cytoskeletal system exists inside prokaryotic cells, and no intracellular movement is observed in live prokaryotes.

In our view (Melnitsky and Margulis 2004), the universally distributed cytoskeletal system evolved from a primary source. Whole spirochetes, with their genomes, were acquired by symbiosis. We posit that intracellular motility originated from free-living, highly sensitive, helically shaped bacteria, many of whose descendants are easily found in sulfurous muds, in ponds, and in the sea (figure 13.1). Though it sounds like wild

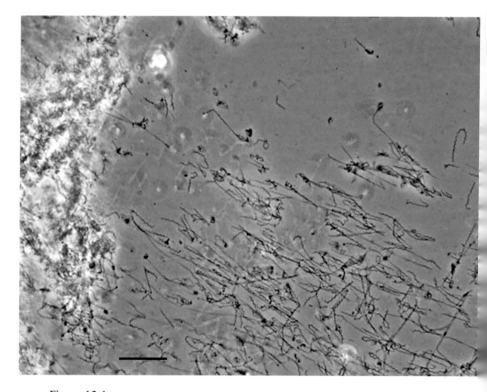

Figure 13.1
Spirochaeta perfilievia, the sulfide-requiring round-body forming spirochete—studied for more than 40 years by Galina Dubinina and her team—that we suspect best resembles the ancestor of the cilia (Wier et al. 2010). Although Hall's (2011) analysis of the spirochete contribution to eukaryotic nuclear genes was forced to use *Leptospira* because the *S. perfilievia* genome is not sequenced, we hypothesize that the anaerobic oxygen-tolerant spirochetes of sulfide muds worldwide are closer than any leptospires to the ancestors of undulipodia. Scale bar = 10 µm.

speculation, we have evidence that spirochete genes and proteins were ancestral to the internal evolution of certain organelles, including the kinetosome-centriole-centrosomes inside many eukaryotes. The earliest eukaryotes, which couldn't even breathe oxygen, were animated. From the early Proterozoic eon, ancestral protoctists moved slowly or vigorously, dependent on environmental conditions. By the end of the story, at the base of the Cenozoic era of the Phanerozoic eon, 65 million years ago, the spirochete precursors to modern organelles had evolved to become, in vertebrate cells, the sensory cilia of our tongues, nasal linings, auditory kinocilia, and retina.

Outside Becomes Inside

As outdoor patios become indoor greenhouse gardens, footbridges are roofed by translucent plastic panels, or skyscrapers are joined by a heated mezzanine bridge, the outside is converted to an expanded inside. Tunnels and roofing bridges beyond the shelter's walls are made to fuse into a network. Soon, new modulated inside spaces provide refuge from winter's blizzards and summer's downpours.

This is one strategy of life. Social insects, fossorial (digger) mammals, and other species literally incorporate patios of their surroundings with inevitably smaller citizen-inhabitants and their potential offspring. Again and again, over more than 3,500 million years, what once had been outside "the body" became covered with secretions, exudates of interacting local now-permanent residents. Bodies of different origins and proclivities joined to form vibrant communities that become new units of life, new "individuals" at larger, more inclusive levels of organization.

Such living together of organisms of different types, strains, or species in protracted physical association is called symbiosis. The German botanist Anton de Bary coined the word in 1873 from Greek "sym" (= with) and "bios" (= living). He called the partners—organisms with different names that live together in the same place for protracted periods—symbionts. Therefore, the word "symbiosis" refers to an ecological condition: the presence of one type of organism becomes the persistent environment of the other. By 1907, Konstantin Sergeevich Mereschkovsky (see chapters 8 and 12) had rejected Charles Darwin's concept of natural selection as the main source of new traits in evolution. Mereschkovsky could not fathom how a process that merely eliminates members of populations (by death or by failure to survive or to reproduce) could have generated the diverse and prodigious life forms that he studied: pond-water ciliates, green hydras, and marine algae. He surmised that many of the innovations he saw in the natural world were the results of symbiogenetic mergers. Not only was Mereschkovsky correct; by 1921 his young compatriot Boris Mikhailovich Kozo-Polyansky had related his "symbiogenesis" to Darwin's natural selection in a fully modern way that explained the major features of this evolutionary process. Kozo-Polyansky's book, a brilliant, astounding, and insightful review of the literature titled *Symbiogenesis: A New Principle of Evolution* (1924), was ridiculed, denigrated, but mostly ignored in all languages, even Russian, for more than eighty years.

As the word "chimera" suggests, all organisms visible to the unaided eye are integrated former symbionts. Evolving from ancestral units, organisms of different kinds lived together until they periodically or permanently merged and covered themselves with exudates, skins, sheaths, or coats made by both partners. Over evolutionary time, the outside became the inside at a higher, more inclusive level of living organization.

Two Kinds

Most of life on Earth is recognizably bacterial in structure (i.e., cells are prokaryotic; they lack a nucleus, protein-studded chromosomes, and intracellular motility visible in the live being). Sexuality is unidirectional: one partner (the donor) passes genes to the other (the recipient). All non-bacterial (non-prokaryotic) life forms are nucleated organisms, composed of eukaryotic cells. Not only does each cell have at least one nucleus; each has an ancestry ultimately derived from integrated co-evolved bacterial communities. The nucleus originated tethered to the microtubule intracellular motility system. Itself a legacy from an ancient organellar system that has extant descendants, the nucleus was released from an organellar complex called the karyomastigont in a series of events too complicated to explain here (Dolan et al. 2004; Margulis et al. 2006). These events correlate with the long-term presence of a sulfide-rich ocean during the Proterozoic eon (Margulis et al. 2005). All cells of protoctists, fungi, animals, and plants have multiple ancestry: they evolved from integration of identified bacterial components in a specific sequence (see appendix A; also see Margulis and Chapman 2010). Nucleated cells, all of them, are integrated growing populations of smaller microbes. They represent communities of bacteria that, since they originated more than 1,000 million years ago in the Proterozoic eon, have propagated as units, often called monads. These nucleated units, eukaryotic cells as evolutionary lineages, have never entirely disintegrated since their inception in the middle of the Proterozoic eon (Margulis et al. 2006).

Kinetosomes, Centrioles, and Centrosomes

Three intracellular complexes common to virtually all eukaryotic cells show tantalizing evidence of a common origin. Our best guess is that this origin involves the genetic contribution from one of their earliest bacterial ancestors: free-swimming spirochetes. Kinetosomes, the basal structure with a peculiar ninefold symmetrical structure that underlies waving and undulating cell "hairs" (e.g., cilia and sperm tails), are identical to centrioles, the little cylinders, composed of microtubules, that sit at the poles during mitosis (cell division) in animal cells and in many protoctist cells. The only difference is that centrioles lack the exterior whip part of the sperm tail. In many egg cells and other animal cells, the centrioles are embedded in larger ball-shaped structures replete with many microtubules. They are called centrosomes. That kinetosomes, centrioles, and centrosomes are related is obvious to anyone who has watched or stained these cell organelles.

The minimal nucleated cell represents at least two formerly free-living bacteria that differed greatly. One was an archaebacterium ("Archaea") and the other a type of eubacterium. (See appendix A.) We don't know whether any extant free-swimming descendants limited to only these two ancestral microbes still exist. We have not yet found them. But hundreds if not thousands of extant eukaryotes descended from more than these two types of merged microbes (spirochetes and archaebacteria) before the symbiotic acquisition of the alpha proteobacterium, the O_2-respiring bacterium that became the mitochondrion.

Undulipodia from Spirochetes?

A coherent series of widely distributed enzymes and structures evolved from common ancestors. The scientific literature abounds with evidence that the sensory systems of humans and all other mammals—which recognize and respond to touch (mechanoreception), to light (photoreception), to sound waves (auditory sensitivity is also based on mechano reception), and to chemical sensory signals (smell and taste)—evolved by redeployment of undulipodia. (See plate VII.) Olfactory, gustatory, and other chemical perception is, fundamentally, chemoreception. Gravity receptors are specialized mechanoreceptors that sense dense minerals such as calcium phosphate or calcium carbonate "stones" (statoliths) inside cells. Just as magnetotactic bacteria swim toward magnetic minerals, such as iron oxides (Fe_3O_4–Fe_2O_3), we know that magnetotaxis evolved first in motile bacteria. In the same way we suspect that touch, chemical attraction (taste, smell), and responses to light and gravity evolved in swimming bacteria whose descendants still react immediately to these stimuli.

Origin of Mitosis and Protoctists

In many nucleated cells (e.g., *Trichomonas* and other parabasalids, green algae such as *Chlamydomonas*, golden-yellow algae such as *Ochromonas*), and in the male gametes (sperm) of mosses, of mammals, and of many other animals, the haploid system is conspicuous throughout the life history of the growing and expanding cell. Just before cell division, the centrioles, with their peculiar ninefold symmetrical protein structure at the base of each undulipodium (abbreviated [9(3)+0]), reproduce, and two centrioles are detected where one preceded it. In the offspring, a thick bundle of microtubules forms the mitotic spindle, which then elongates. The result is a double set of organelles that appear again, often at the anterior ends of the cells. Later, a doubled set of centrioles form by division of the kinetosome. Although no nucleic acid synthesis of any sort of DNA has been detected inside the peculiar structure, a special DNA in the nucleus codes for the centriole-specific RNA (Alliegro and Alliegro 2008; Alliegro et al. 2010). Special RNA synthesis is intimately associated with reproduction and growth of new kinetosome-centrioles. Kinetosome-centriole-associated structures include asters, centrosomes, pericentriolar material, rhizoplasts (nuclear connectors), axostyles, and parabasals. Made of homologous proteins, they are seen, in one form or another, in virtually all nucleated cells. We interpret this reproduction-growth cycle of the cytoskeleton, the intracellular motility cytological system, as a relict of the original sulfide-oxidizing motile spirochete bacteria from which the undulipodia were derived (Wier et al. 2010). The archaebacterium and the eubacterium from which the eukaryotic cell formed produced the mitotic spindle and the various kinds of cilia (including vertebrate sensory cilia) and sperm tails. The generic name for this distinctive organelle is "undulipodium." We hypothesize that the undulipodium evolved from a common spirochete ancestor. Its descendants extant today include *Treponema, Borrelia, Spirochaeta per-filievia* (Dubinina et al. 2010), and all undulipodia.

Origin of Meiotic Sex in Protoctists

The kinetosome-centriole itself from which the proteinaceous motile shaft [9(2)+2] axoneme began was the spirochete attachment site analogous to that shown in figure 13.1; DNA of this eubacterium, the DNA of the original partner, underwent conjugation with its archaebacterium partner in the formation of the nucleus (Margulis et al. 2005, 2006). The co-evolution of the two genomes led to development of the cytoskeletal system, including the spindles of microtubules. The failure to understand the crucial evolutionary processes that involve cell motility leads to egregious errors of interpretation of "family tree" diagrams, especially in the "molecular evolution" community (Dolan 2005). With evidence for a spirochete genetic contribution to the genomes of eukaryotes obtained from samples of at least ten diverse eukaryotes, ranging from anaerobic protists to mammals, fungi, and plants, Hall (2010) performed a rigorous molecular analysis. His methods permit interpretation of complete protein sequence data (proteomics). Hall's evidence shows unequivocally that ancestral spirochetes contributed the genes for at least sixty proteins to the eukaryotic nuclei of the organisms sampled. The simple branching trees in the literature of genomics and proteomics literature are intellectually flawed (Sapp 2009).

A burgeoning literature supports the view that the major protoctist taxa evolved as integration of symbiotic bacterial partners proceeded, outside became inside, and cells enlarged. The discovery of the scleroderma antigen (a protein conspicuous in a human autoimmune disease) in the "neck" of "Rubberneckia" (*Caduceia versatilis*, a termite symbiont that lacks mitochondria) has been interpreted by Dolan et al. (2004) as a clue to the retention of the ancient evolution (Proterozoic eon) of the cell motility-mitotic system that evolved in protoctist ancestors to animals, fungi, and plants (Margulis et al. 2005).

Sensory Cilia of Vertebrates

As we know from chapters 3–6, the sensory systems of swimming bacteria were already remarkably competent. So were bacterial social and community lives complex. (See chapters 5–8.) Recognition of water that contained various types and quantities of salt and its absence (desiccation), of food as a source of carbon, and of light as source of energy or information supplemented mechanical stimulation (touch, sound) long before the origin and evolution of animals with brains. The specific evolutionary lineage as the base of the eukaryotes, we suggest, is that of the propagule-forming eubacteria, ancestors of modern *Spirochaeta perfilievia* spirochetes (Dubinina et al. 2010; see plate VIII in the present volume). These wrigglers attached to hot, acid-resistant sulfidogens—*Thermoplasma*—as they still associate and attach to sulfide-making bacteria today. From the symbiogenesis, the motile first eukaryote, with its karyomastigont, emerged. Now, with the nucleus, the undulipodia, as part of the karyomastigote, tens of thousands more protoctist species (ciliates, mastigotes, water molds, etc.) evolved. Many plant and animal sperm cells that retained the karyomastigont also eventually evolved. Ciliated sensory cells of animal epidermis that show the 9(2)+2 cross-section at the electron-microscope level form the basis of the animal sensory system, as is detailed in Vinnikov's 1982 book on the detailed structure of vertebrate sense organs.

What communities thrived outside in sulfur-rich anoxic muds became integrated inside the partnerships that evolved into nucleated cells. Although early dynamic duos (which became anaerobic archaeprotists) prevailed over the unassociated spirochetes and unattached archaebacteria, the tiny unhitched prokaryotic associates persisted on their own, too. In temporary associations in tight communities, and by themselves, they still can be seen, unfused and little changed, in muddy, sulfurous habitats that have persisted over 3,000 million years.

14

Packaging DNA

Andrew Maniotis

Many of the most fascinating and evolutionarily momentous changes in life's history took place at a level below the resolution of the unaided human eye, and before the evolution of any animal. Here Maniotis traces the transition from the bacterial carrier of DNA, a circle of genes technically known as a genophore, to the chromosomes within the more complex eukaryotic cells, such as our own.

While mountain ranges, volcanoes and their craters and lakes, coral-bivalve-algal carbonate reef complexes, and tropical forests come and go, oxygenic blue-green bacteria (cyanobacteria) look, act, and probably taste just as they did 2,000 million years ago. Of course, mutation and other heritable changes are both selected against and selected for, with consequences that perpetuate genomes. But I suggest that the contiguous chromosome-chromolinker genetic system that is characteristic of nucleated organisms—protoctists, animals, fungi, and plants—evolved from the expansion and folding of early prokaryotic genophores. Especially after eukaryotic chromosomal genomes replaced the prokaryotic chromonemal ones, life overcame a continuous harsh entropic onslaught from its abiotic universe.

How has the genophoric-genomic system, composed of nucleotide sequences of DNA and/or RNA, communicated with the universal sensory system of so many diverse organisms? The cells of life sense and respond to energy flow in both the internal and the external environment. Incessant metabolism-mediated energy flow, whether mechanical, chemical, magnetic, or photic, is a *sine qua non* of life.

Extensive study indicates that stress (starvation, water scarcity, chemical toxicity, excessive light, heat, cold, acidity, salt, crowding, and so on) induces specific responses: syntheses of small signal compounds (known by such names as "pheromones," "hormones," "allelochemicals," "semiochemicals," and "alarmones") that interact with the RNA or DNA genome.

With microsurgery, we have removed all the chromosomal genes from the living cells of animals during the process of cell division. Once outside the cells, these complete chromosome sets retain the ability to undergo dynamic structural transformations in response to changes in salt concentrations, in protein solutions, or in polyamines. The isolated chromosome sets can be induced to disassemble and reassemble under microscopic observation because a thin thread of DNA physically connects all chromosomes of a set. In all our experiments, mitotic genomes of eukaryotic organisms are removed as wreath-like structures. After proteolysis and re-addition of histones, topoisomerases, and other proteins, they re-form their wreath-like structures, in which the chromosomes remain linked together by threads of DNA we call chromolinkers. Proteins along threads associated with the chromosomes are elastic, whereas chromosomal and interchromosomal DNA behaves as a continuous rigid wire (figure 14.1).

In living cells, when we remove the set of chromosomes (the genomes), we discover giant wreath-shaped chains that have, along their length, folded regions that we recognize as mitotic chromosomes (Maniotis et al. 1997). The chromosomal wreath may explain how, since the chromosomes are joined, they typically segregate with precision. After DNA replication and condensation into chromosomes during prophase, chromosomes are not physically independent; rather, they are folded regions

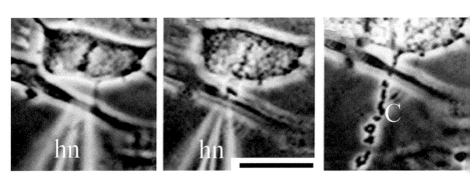

Figure 14.1
A chromosome is harpooned out of the metaphase plate of a mammalian endothelial cell in mitosis, and tension is applied. All chromosomes succumb. The fact that all are pulled away together demonstrates that the nuclear genome is a single, physically connected, system even in metaphase. hn: harpooning needle. C: chromosome set as single linear structure. Scale bar = 30 μm.

along two separate yet continuous cables that pull apart during anaphase after they untangle and as they maximally condense. These topological transitions may also account for why chromosome condensation and interchromosomal tension are required for the metaphase-to-anaphase transition, called "checkpoint control," that determines the timing of chromosomal segregation in anaphase.

The remarkable fidelity of chromosomal behavior in cells results from the evolutionary history of genomes, in my opinion. A pair of giant wreath-like structures in diploid animal cells evolved from augmentation by protein and folding of an ancestral prokaryotic circular genophore. Viewed this way, the diploid mitotic nucleus of animal and plant cells now comprises two intertwined chains, each representing a haploid chromosome set. They first move toward the center of the cell because of growth and shrinkage of compression-resistant microtubules. Then the two half-genomes simply pull apart as two complete loops, not as 46 independent, separable chromosomes. A continuous elastic lattice interspersed with discontinuous compression-resistant struts conforms to structural and mechanical requirements of tensegrity (tensional integrity). This architectural design principle leads to efficiency and strength with a minimum of material (figure 14.2).

Chromosomes appear as discrete, countable units in fixed and stained preparations of animal and plant cells in mitosis (Manuelidis 1990; Nagele et al. 2001). As Mendel's work was rediscovered, and as the chromosomal basis of inheritance was developed during the first several decades of the twentieth century, primarily using the fruit fly *Drosophila*

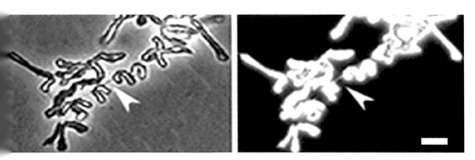

Figure 14.2
Chromolinkers, strands of thin connection between the individual chromosomes in a cell. Here they connect the diploid chromosome set from newt cells (left). Freshly extracted newt chromosomes chromolinkers connect similar-size chromosomes (arrow). Fluorescent image (right) of set of chromosomes stained with ethidium bromide shows that chromolinkers are composed of DNA. Scale bar = 5 µm.

sp., chromosomal physical/mechanical separateness and random distribution among chromosomes was asserted to provide the physical-chemical basis for meiotic recombination, prometaphase chromosomal movements and arrangements, chromosome movements at the mitotic cell equator, checkpoint control, and the metaphase-anaphase transition. In contrast, bacterial genophores, as large or small replicons (e.g., genophores or plasmids, respectively), typically are depicted as continuous ring-like structures that, through semi-conservative base-pairing, produce a new ring-like structure along the pre-existing genophore or plasmid. When replication of the DNA loop is completed by attachment of a membranous "mesosome" at some point along each of the replicated loops, the replicated loops are separated as the parent bacteria elongate to form the two offspring cells. It was established long ago that dinomastigote protists have a single, long, condensed chromosome-like genome, deficient in standard histone chromosomal proteins, whereas certain ciliates (e.g., *Euplotes* sp.) replicate "U-shaped" genomes in which RNA synthesis and DNA synthesis are spatially separated as a band of replicating DNA moves around the "U" until synthesis is complete.

The perception of chromosomes as physically separate countable entities was suggested when meiotic chromosomes extracted from some insect cells were observed to behave as individual units after detachment from the meiotic spindle and manipulation in the living cell. In textbooks on genetics, physical discontinuity between chromosomes is taught. Results from experiments in which insect cells were micromanipulated are generalized to all cells (Glanz 1997). Other types of experiments, in which chemically extracted DNA molecules were isolated from populations of mitotic *Drosophila* sp. cells, showed that shearing in a horizontal rotating viscometer (used to measure viscosity of a fluid) generated DNA size values that were similar to the estimated sizes of the largest metaphase, anaphase, or G1 chromosome. These viscometry measurements were derived from biochemically extracted and physically sheared DNA. Despite questions about actual DNA molecule lengths in certain insect cells that contradicted the viscometry measurements of sheared chromosomes, misinterpretations became accepted and established. The idea that the longest single DNA molecule in living cells was at least as long as the longest chromosome, but not longer, became dogmatically accepted.

Similarly, the idea of "mitotic checkpoint control" assumes that each chromosome acts separately and independently of the others. The results of our work contradict both of these preconceived ideas. We, and others, have shown that mechanical tension among continuously attached chromosomes is the essential signal for the sets of chromosomes to move in a coordinated way as "mitotic plates."

The concept of an interconnected genome has precedents. McClendon (1901) attempted to isolate individual chromosomes from *Cerebratulus* eggs so as to transfer them into another cell. He could not. When he tried to remove a single chromosome, all the other chromosomes were pulled out of the cell in succession, as if they were connected to the one he manipulated. Because chromosomal discontinuity has never been seen in mammalian, bovine, avian, or amphibian genomes either during or after extraction from any living mitotic cell, we suspect that the apparent physical separateness of the chromosomes observed in (for example) insect meiotic genomes during micromanipulation may have been observed because the exceedingly fine interchromosomal linking elements (chromolinkers) were inadvertently broken during the manipulation, or during the shearing of the DNA. Our assertion regarding the possibility of chromolinker breakage during manipulation of chromosomes in insect cells was made more likely by the fact that Bruce Nicklas's group at Duke University in North Carolina, which studied crane-fly spermatocytes, later reported that the physical linkage between chromosomes influenced the positions of other chromosomes during manipulation (Li and Nicklas 1995). For example, Li and Nicklas "rescued" meiotic cells blocked in mitotic metaphase due to chromosomal misalignment by moving only one chromosome, which suggests that chromosomal positions are physically communicated to all the other chromosomes (Li and Nicklas 1995). Their observations are consistent with our chromolinker concept, according to which all chromosomes of the genome are connected.

We accept these and several other results as compelling evidence for a non-random organization of mammalian chromosomes. What is the significance of interchromosomal continuity for physiology? What are the implications for understanding the evolution of the eukaryotic genome from earlier, smaller, and simpler genomes of bacteria?

The apparent physical discontinuity of animal chromosomes is due primarily to harsh chemical and physical treatments during isolation, when chromosomes are increasingly separated and rendered rigid and separate. The random positions of the individual chromosomes during routine chromosome fixation and staining (preparation for detailed microscopic study) is likely to be caused first by chemical (colchicine) collapse of the mitotic spindle, salt shock, or squashing and other harsh routine treatment. The relative positions of chromosomes are distorted during the preparative steps, and the chromolinkers are broken, especially during dehydration.

In contrast to this picture of chromosomal separateness during routine preparative technique, physical removal of genomes from living mitotic cells under physiological conditions tells us that the genomes of eukaryotic metaphase cells are structurally continuous and highly ordered. If metaphase chromosomes in particular (because they are the most highly coiled) are discrete and physically disconnected from one another in the living cell, one would expect that occasionally some individual chromosomes would remain in the cells, or at least that individual chromosomes sometimes would be extracted and leave others behind during micromanipulation. But in thousands of isolation experiments, physical extraction from a variety of cells has never yielded a single individual chromosome in our lab or in any of many other labs.

The continuity of the chromatin, the protein-DNA complex of human and other mammalian cells, is shown in figure 14.1. The physicochemical basis for the inheritance of the human sensory system (touch, hearing, taste and smell, vision) is coded in the DNA of the genome in this photograph of the united chromosomes. Only in an adequate environment —a cellular, whole body, social and community ambience—does this DNA behave to help generate a sensitive, sensible developing person. Fantastic as it may seem, basic aspects of the chromosomes that biochemically define who we are evolved more than 3,000 million years ago in the genophores of the responsive, wily bodies of our mute bacterial ancestors.

Plate I Second Connecticut Lake, Pittsburg, New Hampshire. View to the north from the central western portion of the lake. The nodule field begins 4 meters from the shore at a water depth of 4–8 meters.

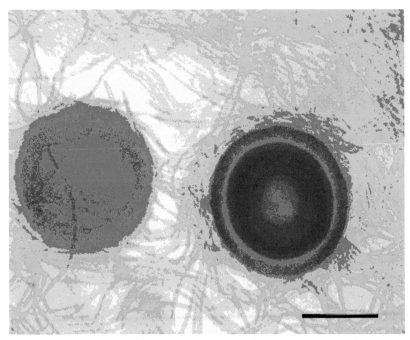

Plate II Left: Immature ooid formed by at least ten different microorganisms that include *Phormidium* sp., green filamentous photoautotrophic oxygenic bacteria. Right: Layered calcified sphere that becomes an ooid, a mature carbonate sand grain. Scale bar = 30 μm.

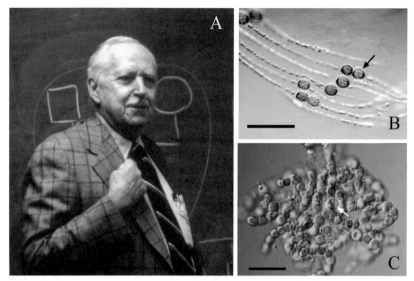

Plate III (A) Murray Bowen explains his theory. (B) Bacterial altruism. Heterocysts, specialized cells (at arrows), fix nitrogen gas from the air to produce organic nitrogen compounds and then they die. The nitrogen compounds are quickly passed to neighboring reproductive cells. (C) *Nostoc* sp. cyanobacterial filaments after removal from the plant and growth in culture. In nature these nitrogen-fixing cyanobacteria are required endosymbionts that live inside the *Gunnera manicata* plant cells. Scale bars = 10 μm.

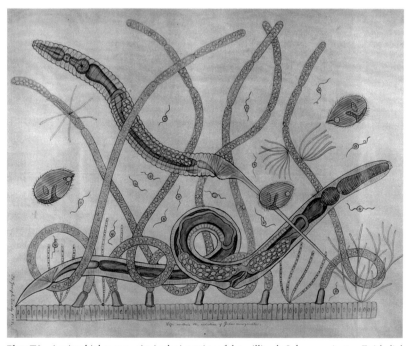

Plate IV A microbial community in the intestine of the millipede *Julus marginatus*. Epithelial cells of the arthropod at the bottom to which are attached six trichomyces fungi. Two different species of nematodes packed with eggs are shown among swimming mastigotes and ciliates. Also attached to the intestine and to trichomyces is the anthrax bacillus *Arthromitus* sp., both branched and unbranched, with and without its spores. Drawing by Joseph Leidy, 1888.

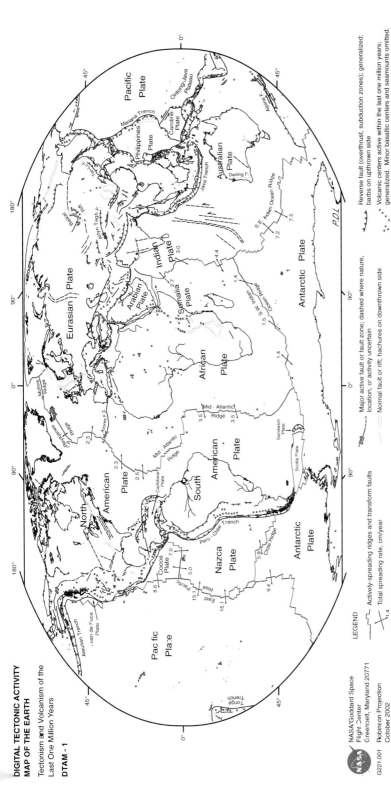

Plate V The Digital Tectonic Activity Map (DTAM), a digital atlas of the tectonism and volcanism of the end of the Pleistocene period (the last million years), was compiled from field, geophysical, and satellite measurements. The DTAM is a unique tool for understanding the physiographic nature of our dynamic Earth. Major active faults or fault zones are shown by solid lines. Dashed lines show where the location or degree of activity is not precisely known. Extensional ("normal") faults, shown in gold, have hachures on the downthrown side. Compressional ("reverse") faults are shown with barbs on the upthrown side. Currently active volcanic centers, known from historical records or geomorphic freshness, are shown as red dots.

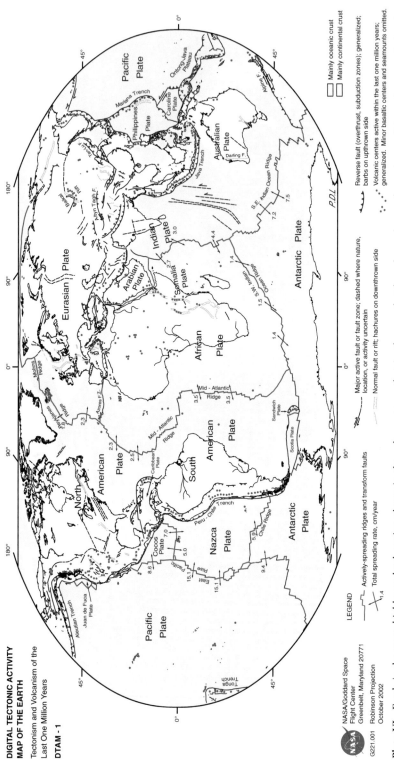

DIGITAL TECTONIC ACTIVITY MAP OF THE EARTH

Tectonism and Volcanism of the Last One Million Years

DTAM - 1

NASA/Goddard Space Flight Center
Greenbelt, Maryland 20771

G221.001 Robinson Projection
October 2002

LEGEND

Actively-spreading ridges and transform faults

Total spreading rate, cm/year

Major active fault or fault zone: dashed where nature, location, or activity uncertain

Normal fault or rift; hachures on downthrown side

Reverse fault (overthrust, subduction zones); generalized; barbs on upthrown side

Volcanic centers active within the last one million years; generalized. Minor basaltic centers and seamounts omitted.

Mainly oceanic crust

Mainly continental crust

Plate VI Earth is the most highly evolved, active planet in the solar system—that is, the one that has changed most from the conditions that prevailed at its formation. This schematic global tectonic activity map (GTAM) is intended to give a realistic picture of tectonic activity "at present." We include here the past million years—long enough to be representative of the Pleistocene epoch, but short enough that modern landforms produced are recognizable. Global tectonics is a view of the structural processes and motion of Earth's crust. Plate tectonics is the theory that Earth's lithosphere, consisting of the crust and the upper mantle, is divided into discrete rigid plates that are in motion with respect to one another.

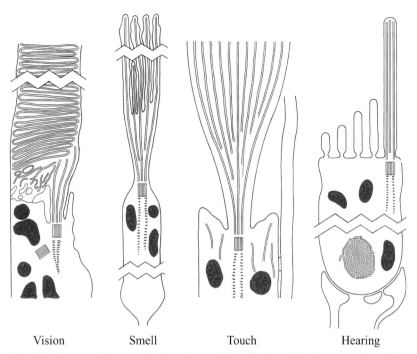

| Vision | Smell | Touch | Hearing |

Plate VII   Sensory cilia, examples of products of diversification of the kinetosome/centriole-undulipodial system that persists in modern vertebrates as sensory cell processes for smell (olfaction), balance and mechanoreception (touch), and vision (retinal rods) in animals.

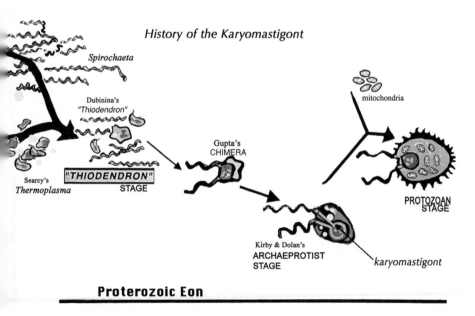

Plate VIII   Eukaryosis: origin of the karyomastigont (the nucleo-cytoskeletal system), summarized in Margulis et al. 2006. Also see Sapp 2009.

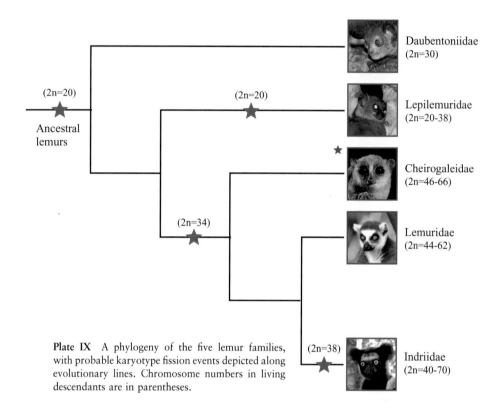

Plate IX A phylogeny of the five lemur families, with probable karyotype fission events depicted along evolutionary lines. Chromosome numbers in living descendants are in parentheses.

Daubentoniidae (2n=30)

Lepilemuridae (2n=20-38)

Cheirogaleidae (2n=46-66)

Lemuridae (2n=44-62)

Indriidae (2n=40-70)

(2n=20) Ancestral lemurs

(2n=20)

(2n=34)

(2n=38)

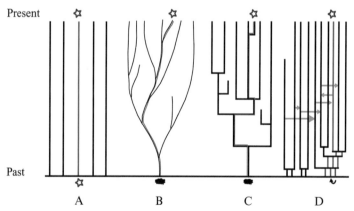

Present

Past

A   B   C   D

Plate X These diagrams show the four concepts that explain how the starfish or any animal we see evolved from its different ancestor. (A) No evolution. Starfish of the past were created, 6000 years ago, in the same form as the living starfish. (B) Evolution by descent with modification (Darwin 1859). (C) Punctuated equilibrium (Eldredge and Gould 1972). Many new species evolve rapidly after mass extinction. Then the rate of speciation declines and a "candelabrum" topology is generated. (D) Larval transfer (Williamson 2011). The starfish evolved from different ancestors which hybridized; in sequence they produced the larvae and then the adult.

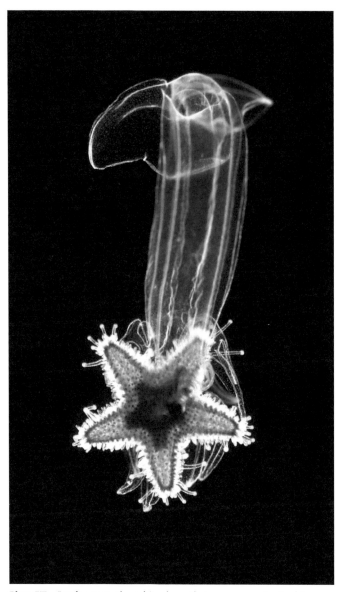

Plate XI   *Luidia sarsi*, the echinoderm that announces its double parentage: the single fertilized egg contains two different animal genomes of separate origin. The larva, the translucent bag at the top of the photograph, hatches first from the egg; the tiny starfish adult then develops inside the bag. The starfish is released and settles to begin life in the benthos. Most echinoderm larvae develop into radially symmetrical adults, but in *L. sarsi* both the bag-like larva and the starfish adult independently coexist for as long as 3 months. The fact that a single fertilized egg develops into two different organisms is explained as hybridization followed by sequential gene expression. The ancestors of the larva (that was an adult in another lineage) hybridized in the evolutionary past, probably at beginning of the Mesozoic Era, more than 200 million years ago.

Plate XII A bobtail squid (*Euprymna scolopes*) that nightly demonstrates belly-side counterillumination. The light emanates from millions of bioluminescent symbiotic bacteria that are daily replenished, maintained, and grown inside the animal in paired ciliated light organs.

15

Lemurs and Split Chromosomes

Robin Kolnicki

Despite repeated claims, there is very little evidence for gradual transition from one to another species. Here Kolnicki explains one of several proposed methods for rapid evolution: "karyotypic fissioning," also called "centromere-kinetochore replication theory." This theory focuses on the behavior of chromosomes and illuminates the history of a branch on our own lineage. Our 50-million-year-old primate relatives evolved into more than thirty species of monkey-like animals when, because of plate tectonics and spontaneous changes in their chromosomes, they became isolated on the great island Madagascar.

Lemurs of Madagascar

Madagascar, the fourth-largest island in the world, is located in the Indian Ocean off the southeast coast of Africa. It is home to a biota that is unique in all of life's kingdoms. Many of its animals, chronicled in fables, are indeed fabulous. These include many species of lemurs that are found nowhere else in the world. They share characteristics not found among the other primates, such as female dominance and greater dependence on olfaction (sense of smell) over sight. Unlike nearly all other primates, lemurs display cathemerality (see chapter 22). It distinguishes them from nocturnal and diurnal mammals. Lemurs actively forage and socialize during portions of both day and night.

Lemurs, today classified in five families, were among the first suborders to diverge in the order Primata of our vertebrate class, Mammalia. This order includes lemurs, lorises, monkeys, orangutans, bonobos and other chimpanzees, humans, and other apes. Lemurs may have colonized Madagascar as early as 80 million years ago or as late as 50 million years ago. Subfossil (recently extinct) lemurs found to have lived as late as 500 years ago were larger than the largest extant lemurs and the largest humans. Their recent extinction was probably influenced by hunting and habitat destruction. Today lemurs are endangered primarily because their natural habitats are becoming increasingly fragmented and diminished in size by slash-and-burn agriculture, livestock grazing, and logging.

Studies of the anatomy and the physical structure of fossil and living lemurs, along with behavioral and genetic analyses, have partly illuminated their evolutionary history. The fossil record alone does not help to determine the dates of divergence, because subfossil lemurs are part of the diverse contemporary fauna and not predecessors to modern lemurs. Some of the best evidence for the timing of divergence of lemur families, therefore, comes from genetic studies. Del Pero et al. (2006) used mitochondrial and nuclear DNA to investigate relationships among living lemur taxa, and generated an acceptable phylogeny. (See plate IX.) Although this family tree depicts reliable evolutionary relationship between the families, the processes that preceded diversification of lemurs from ancestral primates are still unknown.

Speciation, some involving divergence from one ancestral lineage into multiple descendant lineages, can result from various natural events, including the formation of geographical barriers. Geographical barriers (e.g., waterways, new mountains, volcanoes) lead to allopatric speciation, a gradual process whereby interbreeding is prevented when one population is physically separated into two or more populations. One instance of allopatric speciation in lemurs occurred when the Betsiboka River formed a barrier to interbreeding and led to lineage branching in *Eulemur*, *Lepilemur*, and *Propithecus* species (Pastorini et al. 2003). Karyotypic fission is an alternative mode of speciation in mammals that occurs in sympatric populations (that is, within an interbreeding community) with no need for geographic isolation (Guisto and Margulis 1981).

Karyotypic Fission: Neocentromere Formation

Karyotypic fission is a cell process whereby each chromosome in a gamete (sperm or egg) fragments at the centromere/kinetochore region. The entire genetic content of the sperm or egg cell is inherited by the offspring cells, but it is packaged very differently than in the original parent. Large and few chromosomes give rise to smaller and more numerous ones. In my example (figure 15.1), one large chromosome generates four smaller ones. The macromolecular sequences in the region of the chromosome are composed of DNA (centromere) and protein (kinetochore) sequences, respectively. The spindle fibers (bundles of

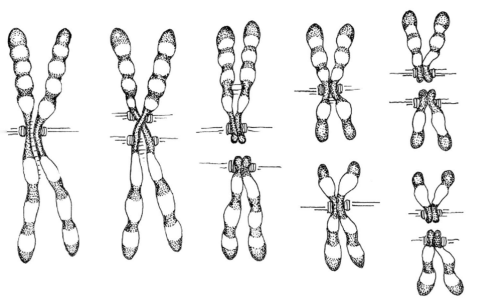

Figure 15.1
Karyotypic fissions generate four functional descendants from a single ancestral chromosome through pre-duplication of the centromere/kinetochore components (genetic-DNA
and protein-amino acid portions) of a chromosome.

microtubules) attach to the centromere/kinetochores during cell division.
centromere/kinetochore is responsible for the proper segregation of chromosomes during mitotic cell division. Chromosomes move along the
spindle fibers toward the poles of the cell. They are attached by their
dynamic centromere/kinetochores to the microtubules.

Gametes contain only half of the full complement of chromosomes.
Unlike body (somatic) cells, gametes (egg and sperm cells) undergo two
divisions: meiosis I and meiosis II. The first division separates chromosome pairs and reduces the chromosome number by half. The second, a
mitotic division, separates sister chromatids. The result is four cells, each
of which contains one copy of each chromosome. In ova, only one of
these offspring cells is a functional gamete, an egg. The other three,
significantly smaller than the egg, become non-functional polar bodies
and disintegrate. Chromosome fission that occurs during the second
division results in the separation of the chromosome halves. The molecular motors at the centromere/kinetochore region attach to the spindle
molecules. Force is generated at these attachments. Chromosomes that
possess the centromere/kinetochore at or close to their middles are called

metacentrics. Karyotypic fissioning that generates two smaller shorter, fissioned chromosomes with centromere/kinetochores near or at the end are called acrocentrics. If all metacentrics of a cell were to divide simultaneously, their offspring cells would inherit twice the number of chromosomes as in the parental cell. Matings shortly after the karyotypic fission events between individuals with ancestral large metacentric chromosomes and individuals with a fissioned chromosome set are fertile.

An individual's complete chromosome arrangement is easily visualized after representative cells are stained and photographed; the preparation is called a karyotype. Variations in karyotypes (variations in chromosome shapes and numbers) result from matings between individuals that have fissioned chromosomes and individuals that have retained the ancestral karyotype. Some offspring inherit whole chromosomes in combination with corresponding fissioned halves. A reduction in fertility is not likely before later heritable changes occur. The variety of inherited chromosomal arrangements is subject to mutation and further modification through natural selection. Certain crosses between unfissioned ancestors and newly fissioned descendants may be more successful than others. They may produce more offspring, which will carry specific new chromosomal arrangements. Other crosses may show reduced fertility, infertility, or other chromosomal incompatibility. This is one way speciation through reproductive isolation within a single population can occur.

Evolution through karyotypic fission has rejected or ignored because of the lack of an explanation of how fissioned chromosomes remain functional after "breaking." My "kinetochore reproduction" hypothesis describes the process by which fission results in viable but shorter chromosomes (Kolnicki 2000; Godfrey and Masters 2000). Duplications of kinetochore/centromeres survive. They are known to generate supernumerary kinetochore/centromere regions on a single chromosome. In fact, many chromosomes have "silent centromere sequences" where former functional centromere/kinetochores or partial centromere/kinetochores developed (Perry et al. 2004). Chromosomes with more centromere/kinetochores are preferentially distributed to functional gametes; those with fewer centromere/kinetochores fail to segregate into the egg and are distributed to the non-functional polar bodies of meiosis in females. In the case of chromosomes split between their duplicated centromere/kinetochore regions, both halves maintain full function for attachment of spindle microtubules. The sticky ends of DNA are repaired by telomerase, an enzyme that is present in gametes.

Centric fission, or the transverse breakage of the centromere/kineto-chore region, also may result in functional fissioned chromosomes (Perry et al. 2004). Cleavage at the centromere/kinetochore region leaves sufficient active sequences of centromeric protein and centromeric DNA attached to chromosome halves. The resulting chromosome fragments segregate normally. As with pre-duplication of kinetochore/centromeres, correct alignment with homologous regions on whole chromosomes during cell division is retained. Variations in karyotype can be maintained through backcrosses within the population.

Fission Driving Evolution

Neil Todd (2000) suggests that the variation of diploid numbers in mammals, such as in families of artiodactyls and in canids, resulted from karyotypic fission. The evolution of mammals does indeed follow a pattern, which is observable in the fossil record. Branching events (appearances of new taxa of animals) correlate well with changes in chromosomal complements. When major lineages diverged, chromosomal arrangements were modified in a manner predictable by application of karyotypic fission theory. A systematic increase in the number of chromosomes where a single long one was replaced by two shorter homologous ones occurred as mammalian lineages became more specialized. Todd hypothesizes that the original chromosomal arrangement consisted of mammals with fourteen large metacentric chromosomes. Among living mammals, the range in diploid number is $2n = 6$–92. Most chromosomes in mammals other than the large X, the sex chromosome, exhibit signs of fission. Retention of the X relative to its fissioned descendants was selected for in reproduction. The long arm of the X chromosome is the same in most if not all placental mammals. Placental mammals tend to have the same large amount of DNA per chromosome. Whether or not the chromosome is one arm with telocentric centromere/kinetochores or two arms attached at the single metacentric centromere/kinetochore does not affect viability or fertility. The overall DNA content in marsupials' X chromosomes is substantially less than that in eutherians. In placental mammals, the X chromosome is about 5 percent of the total genomic DNA.

Chromosome fission probably was important in the evolution of Marsupialia. Both fission and single chromosomal fusions have contributed to the Y sex chromosomes of the kangaroo. In another marsupial, the swamp wallaby, females have standard XX chromosomes, as do humans, but males carry a fissioned Y chromosome in addition to the typical X. (That gives males a karyotype of XY_1Y_2.) The monotremes (platypus and echidna species, ancestors of the earliest mammals), which lay eggs as birds and reptiles do, show both the typical mammalian unfissioned karyotype (the karyotype of the duck-billed platypus is XX/ XY) and presumably the subsequent fissioned pattern (the echidna females have $X_1X_1X_2X_2$, whereas the males have a fissioned X_1X_2Y). The general case remains that in most mammals the sex chromosomes are large and unfissioned whereas all other chromosomes (autosomes) have many different patterns of fission.

In the order Primata, Old World primates (including humans) have undergone at least three separate fission events. The first event, which occurred about 38 million years ago, may underlie the origin of different primate families of Old World monkeys and apes. This initial event generated diploid numbers of $2n = 40+$, where the ancestor had diploid numbers of about $2n = 30+$. A s second fission event, about 20 million years ago, probably was the basis for the diversification of the hominoid group, which includes orangutans, gorillas, chimps, and humans. Diploid numbers increased from low numbers in early mammals to that typical of living primates ($2n = 46–48$) in the four ape groups. A separate, secondary fission event occurred in guenons, vervets, mangabeys, and some other Old World monkeys.

Fission in Lemurs

The diverse karyotypes found in the five living lemur families are explained by four separate fission events. Evidence for fission in evolutionary history can be gathered through application of different molecular methods to karyotypes. Chromosomes from animal groups that still retain the more ancestral condition (fewer and longer metacentric chromosomes) are matched to chromosomes of animal groups that have undergone fission. Using stains to distinguish regions (bands) on chromosomes, the distribution of DNA banding patterns throughout the karyotype are compared. As in putting a puzzle together, chromosome patterns are reconstructed from representational clues to produce the most parsimonious ancestral arrangement.

The location of the centromere/kinetochore on chromosomes is also helpful in revealing clues about the history of fission and even the relative timing of the event. Through evolutionary time, the centromere/kinetochore on acrocentric chromosomes relocates again from the end of the chromosome toward the center (figure 15.1). This probably aids in chromosome stability during duplication and cell division cycles. From the inference that a karyotype in a more recent species of mammal has smaller, more numerous chromosomes with centromere/kinetochores at or near the center relative to a its ancestors we infer that a fission event occurred in the ancestors with the larger chromosomes long ago. Mammals whose body cells have karyotypes with smaller, acrocentric chromosomes are thought to have recently fissioned ancestors.

The five living families of lemurs are Daubentoniidae, Lepilemuridae, Lemuridae, Cheirogaleidae, and Indriidae. Daubentoniidae comprises only one living representative species, the Aye-aye. Daubentoniidae was the first to diverge from a common ancestor and has a diploid number of $2n = 30$. Lepilemuridae diverged second and comprises at least seven representative species. Lepilemurs have diploid numbers of $2n = 20$–38. The Cheirogaleidae, which diverged next, have diploid numbers of $2n = 46$–66, followed by the Lemuridae, with diploid numbers of $2n = 44$–62. The last to diverge were the Indriidae, which display the highest diploid numbers among lemurs: $2n = 40$–70.

Four fission events during the evolutionary history of lemurs explain both the diversity of chromosomal number and arrangement in living species. The ancestral condition probably was $2n = 20$, which is still observed in modern lemurs (lepilemurs). A primary fission event before the diversification of modern lemurs generated a population of individuals carrying a variation of karyotypes. *Daubentonia* sp., which separated from the rest of the lemur families early, display chromosomes that are large and metacentric.

A separate primary event underlies the diversification between basal (earliest ancestors in this lineage) Lepilemurids and the rest of the lemur families. A secondary fission occurred some time before the divergence of the lemurids, the cheirogaleids, and the indriids. The basal population probably had a diploid number of $2n = 34$, the minimum number that could generate chromosomal numbers and arrangement observed in lemurids and cheirogaleids.

Finally, a subsequent fission event underlies diversification of the indriid group from a basal ancestor of around $2n = 40$. Although fission can explain many of the modern karyotypes, multiple post-fission centromere/kinetochore relocations and mutations complicate the analyses. Indriid karyotypes are composed of numerous, short chromosomes. And many closely related species display their chromosomes at different stages of centromere/kinetochore relocation. This increases the difficulty of precisely reconstructing the timing and order of chromosomal change.

However, a pattern can still be observed in the karyotypes across the family.

Karyotypic fission has formed the current chromosome patterns, speciation, and zoogeographic distribution in this lineage. Other, less important modes of evolution have participated, but chromosomal rearrangement via karyotypic fission is most consistent with molecular data. Lineages that diverged earlier maintain lower diploid numbers and longer chromosomes. They have not been subject to large-scale modifications. More recently evolved families of lemurs display unique patterns of increased chromosome number and, simultaneously, decreased chromosome size.

The theory of karyotypic fission explains evolutionary patterns in canids (dogs and cats), equids (horses), monkeys, apes, and artiodactyls (goats, sheep, deer). Chromosomal fission and relocation of centromere/kinetochores are significant to reconstructing the history of these mammalian taxa (Imai et al. 1986).

Might a Male's Fissioned Chromosomes Be Sensed?

Reproductive isolation where genetic incompatibility leads to lack of offspring, or fewer offspring, or less fertile offspring puts selective pressure on populations to diverge. Mating between members of two diverging populations diminishes or ceases. Some mammals may be able to detect, presumably by olfactory clues, the fertility status of their potential mates.

Karyotypic fission affects some gene expression, particularly of genes located at the ends of chromosomes (telomeres). The closer a gene is to a telomere, the higher the probability of genetic mutations. Substitutions, deletions, or insertions of base pairs in DNA alter the gene sequence and consequently change the protein structure. Genes that code for olfactory receptors, membrane proteins, or portions of proteins important for scent perception are located near telomeres in primates. Olfactory receptors in primates (including humans) show significant variation due to single-letter DNA mutation. More than 10,000 genes studied in chimpanzees were found to differ from their homologues in humans by approximately 8 percent. Much of this genetic variation involves genes and their protein products that effect reception of olfactory cues. Observable behaviors indicate that odors (chemical olfactants) differently perceived by these closely related primates have a genetic basis. We have seen that karyotypic fission alters chromosomes, and that the alterations include the generation of new centromere/kinetochores and telomeres. Although there is no evidence that karyotypic fission directly causes genetic and protein changes in primate olfaction, this idea deserves further investigation. Olfaction is significant to many aspects of primate social behavior, including mate selection, which suggests that chromosomal changes in olfactory receptors may contribute to divergence and/ or stabilization of mammalian species.

The idea that the chromosome status of a potential mate's fertility may be sensed invites research into direct correlation between chromosomal arrangements and mating behaviors. Fertility status indeed may be sensed, in general, in animal species, nearly all of which are unable to bypass two-gendered parent sexual reproduction by egg-sperm fertilization. Mammals, in particular, reproduce only through bi-parental sex. In some populations of subterranean naked mole rats (*Spalex ehrenbergi*), female rodents reject males of the same species that have differing chromosomal numbers (Nevo 1999). Although female naked mole rats with high diploid numbers ($2n = 60$) tend to prefer males that have lower chromosome numbers, females with lower chromosome numbers ($2n = 52, 54,$ or 58) prefer males that have the same diploid number as they do. The ability of mammals to assess the reproductive status (sterility, complementary gender, potential successful hybrids) requires detailed investigation. We can expect new work on social behavior in the wild, on variations in karyotypes, on correlations with fossils, and on zoogeographic distribution of taxa to lead to more robust phylogenies. We week a consistent reconstruction of lemur evolution in the Cenozoic era that includes all relevant information: molecular biological, biochemical, paleontological, geographic, behavioral, and (especially) chromosomal (karyotypic).

16

Interspecies Hybrids

Sonya E. Vickers and Donald I. Williamson

*Since Darwin, evolution has been presented as a matter of gradual varia-
tions accumulating in branching lineages that eventually speciate. Here,
however, Williamson and Vickers show that evolution was sometimes
radically inventive, producing striking new animal forms in a single
generation by means of fertilization across would-be species boundaries.
Examples include butterflies and starfish and their larvae.*

Here we outline an idea that could radically alter the way we understand
animal evolution. Fertile sex—egg-sperm fertilization—occurred (and
still does) between members of entirely different species, even different
phyla. The rare but fertile hybrids generated striking new animal life
forms in a geological instant.

Neo-Darwinian evolutionary theory ascribes gradual changes to
mutations (alterations in genetic material). The forms of larvae and their
distribution in the animal kingdom suggest a radically different mode of
evolution that involves sudden changes beyond the gradual accumulation
of mutations. We call this "larval transfer" or "origin of larvae by hybridogenesis." The hypothesis is that larvae, or the genes that specify them,
have been transferred from one animal group to another by cross-species,
cross-genera, and even cross-phyla egg-sperm fertilizations.

Larvae, immature stages in the life histories of many animals, may
differ dramatically from the adults that develop from them. We are
familiar with a few larvae, including the caterpillar larva that spins a
chrysalis from which an adult butterfly emerges. But many other types
of larvae exist, and most of them live in the sea. When a larva differs
entirely from the adult into which it develops (holometaboly), we claim
that larval transfer occurred in the history of that lineage.

Most of the evidence comes from larvae of clams, starfish, and sea
urchins, species for which eggs and sperm cast into the sea may merge

in fertilization. Darwin assumed, as do most people today, that larvae and adults begin as a single individual., since they develop from the same egg. The young then are assumed to gradually become more and more different from the adults into which they will eventually develop. This is the core of the "same-stock theory." But no; we explain that the larval forms of animals and the mature adults into which they eventually develop have different ancestors. The larva began as an adult that developed from a fertile egg in one lineage and interbred with an adult from an entirely different animal lineage. Most such anomalous matings didn't even produce offspring, and most of the offspring they did produce failed to survive to sexual maturity. In rare cases of hybrid survival, the genomes of the hybrid would no doubt be so incompatible that gene expression would lead to lethality and hybrids would tend to be selected against. But a hybrid would require survival of only a few individuals, under strong natural selection, to survive. Some hybrids, in fact, left many viable descendants that survived as doubled and even tripled or quadrupled genomes. How did two or more disparate animal beings in the same body at the same time ensure the indefinite continued survival of the hybrids? By expression of the combined genome in sequence rather than concurrently. The larval genome is envisioned to express itself first in the sequence; then the second, third, or fourth genome would express the adult morphology. Metamorphosis from larva to adult is a legacy of the shift from one to another of unlike ancestors.

Patterns of Change

At the time of Darwin, conventional religious beliefs held that the living forms then alive were unchanged over time. Pine trees had always been pine trees, and starfish had always been starfish. (See plate X.) However, the continued exploration of fossils and the discovery that domesticated animals had changed by selective breeding seeded speculation that life forms had evolved through time. When Darwin (1859) argued that this was indeed the case, his idea was rejected by many.

The Darwinian view of life's history is a tree pattern. The bottom of the central trunk represents the common ancestor. The branches denote subsequent diversity, with gradual change in a lineage. The top of the tree represents the present. Branches that fail to reach the top represent extinct life. Because more than 99 percent of all past life on Earth is estimated to be extinct, most branches fail to reach the top. The main limbs that diverge from the trunk depict diversification. Like everyone else until now, Darwin assumed that larvae evolved gradually from the same ancestry as their corresponding adults by natural selection. This "same-stock" (or "common-ancestor") theory has been adopted without question by virtually all biologists since Darwin, but we reject it.

Darwin's paradigm of a gradually branching tree was followed by the paradigm of "punctuated equilibrium." Eldredge and Gould (1972) pointed out that the fossil record contradicts the "numerous, successive, slight modifications" described by Darwin (1859). In the words of Eldredge and Gould, "a species does not arise gradually by the steady transformation of its ancestors; it appears all at once and fully formed." Small mutations and gradual accumulations of change do not fully account for the sudden transformation and rapid diversification so abundant in the animal fossil record. Diversification and extinction in the punctuated equilibrium pattern explain the fossil record by spurts of activity followed by stasis. Eldredge and Gould's successor to Darwin's tree looks like a candelabrum.

One of us (DIW) first proposed the "larval transfer" hypothesis to resolve inconsistencies in comparisons of larvae with the adults into which they develop. We now claim that all dramatically metamorphosing (metabolous) larvae were transferred from other animals through hybridizations between adults in foreign taxa (Williamson 2001, 2003; Williamson and Vickers 2007). In the laboratory, eggs of a sea squirt (a urochordate) were fertilized with sperm of a sea urchin (an echinoderm) (Williamson 1992). A few fertilized eggs showed a remarkable degree of independence between adults and larvae, consistent with the suggestion that they were separate from the start rather than parts of the same stock. A corollary of larval transfer is that larvae were later additions to animal life histories.

The larval-transfer hypothesis proposes an adult animal and its larvae are a chimera that evolved from the fusion of at least two very different animals. The resulting pattern of life's history should be depicted more as a network than as a tree; the "branches" do not bifurcate again and again, as on real trees; rather, they occasionally fuse. The tree of life suggests that matings between like animals have produced a continuous stream of slight differences. In contrast, the larval-transfer hypothesis contends that an animal's ancestry is occasionally the product of two or often more very different animals. It also provides one possible reason for the sudden changes in Phanerozoic evolution, e.g., for Eldredge and Gould's punctuated equilibrium in the fossil histories of animals.

Larval Transfer and Genome Fusion

The idea of larval transfer is not new. According to Francis Balfour (1880–1881), virtually all larvae were "secondary"—that is, they "have become introduced into the ontogeny of species." Balfour was not clear about the larval sources. His concept of secondary larvae has remained ignored and undeveloped until now.

Many examples are known of evolutionary merger by genome fusion in the natural world. In sexual animals, fertilization routinely fuses genomes, usually between members of the same species. However, representatives of different kingdoms fused their bodies and even their genomes (and still do). All photosynthetic eukaryotes (algae, plants, and plant-like organisms such as lichen and Geosiphon pyriforma) are descendants of mergers (Okamoto and Inouye 2006). The gray-green patches on rocks and trees represent fusions between green algae in the protoctist kingdom (or even cyanobacteria) and specific ascomycetes in the fungus kingdom. The ultimate genome fusion occurred when bacteria that respired atmospheric oxygen merged, perhaps 2,000 million years ago, to produce oxygen-respiring protists that permanently contain mitochondria (Margulis and Sagan 2002). Recently, John Werren of the University of Rochester found another merger between the bacterium *Wolbachia* and the cells of several insect species. The entire bacterial genome was discovered as a chromosome arm in the cells of wasps. The entire genome of *Wolbachia* is passed to future generations. The presence of the bacterial genes changes the course of the insects' evolution. Lateral (horizontal) gene transfer, which has been found to be common in bacteria, is increasingly found in other organisms. The novelty of the larval-transfer hypothesis, therefore, is not the merger of two different genomes, but the fact that in developmental metamorphosis each oversees a separate portion of the animal's life history. A complete "changing of the guard" occurs each generation during metamorphosis. Let us look at some of the evidence for larval transfer, and at overlapping development patterns, and then suggest how we might change our paradigm of evolution.

Butterflies and Moths

The most familiar larvae are the caterpillars of lepidopterans: butterflies and moths. Caterpillars are larvae with three pairs of thoracic legs and a variable number of abdominal prolegs. Expandable "eating machines,"

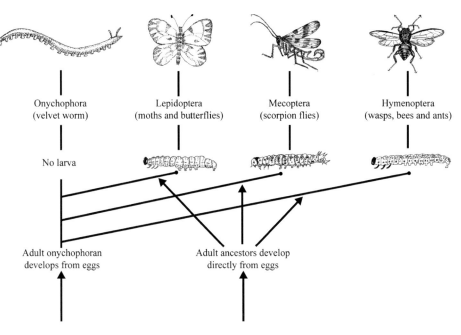

Figure 16.1
The velvet worm has no larval stage and is classified in a separate phylum from worms and insects. Yet strikingly, it resembles the caterpillar the larva in three different orders of insects: butterflies, scorpion flies, and bees. These orders include larval forms that do not develop into flying insects. This discrepancy is explained by our larval-transfer hypothesis: Some, but not all, insects in these orders hybridized with an onycophoran (velvet worm) in the past. The larvae, because they are recent additions to the life histories, are not related directly to the adults into which they develop.

caterpillars can crawl but cannot fly. Caterpillar larvae, however, are not confined to lepidopterans. They also occur in scorpionflies (order Mecoptera) and in wood wasps and sawflies (order Hymenoptera). (See figure 16.1.) Other hymenopterans, including ants, bees, and wasps, have legless grubs as larvae. The occurrence of many types of insect larvae does not correlate with the classification of the adults into which they develop. The larvae and the adults are explicable if larvae are seen as later, relatively independent additions to life histories, as we suggest. Incongruous larvae, inconsistent with the "common ancestral stock" assumption, are illustrated by caterpillars. The caterpillar larvae, so different from moths and butterflies, probably were transferred from adults resembling extant velvet worms of the genus *Peripatus*, found in the organic-rich soils of tropical America, South Africa, and Australia

(Williamson 2009). These worms with legs belong to the phylum ony-cophora and are entirely different from insects or earthworms. If hybrid-ization between various insects and velvet worms occurred, the surviving chimera would have enjoyed the best of both worlds: a larval form spe-cialized for feeding and a flying adult adept at spreading its genes. Both Onycophorans and flying insects have survived rigorous natural selection for millions of years, yet hybridization generated a novelty not available to either ancestor of the chimera.

Chance fertilization between unlike animals probably is infrequent, but it has occurred. Unfertilized insect eggs are occasionally spawned by sperm of different species. Onycophoran fertilization is also external. In the sea, eggs and sperm are often cast into open water, where fertiliza-tions between unlike animals seem more probable. Most of the evidence for the larval-transfer hypothesis comes from marine animals. Some closely related adults have different larvae, and some distantly related adults have similar larvae. The typical idea of evolution from a single common ancestral lineage fails to explain these inconsistencies.

Mollusks and Annelids

Mollusca is a large phylum that includes clams, snails, octopuses, and squid. Annelida, another large phylum of segmented worms, include polychaetes and earthworms. Members of these two phyla are as dis-tantly related as rabbits and butterflies. Most clams and sea snails develop from small translucent trochophore larvae. They share no morphological traits with the adults into which they grow. Octopuses and squid are mollusks that entirely lack larvae. Many polychaete worms, although in a different phylum, develop from trochophore larvae similar to those of clams and snails. Earthworms in the same phylum as the polychaete lack larvae. Some but not all members of several less-well-known marine phyla, including members of Sipuncula, also have trochophore larvae. Octopuses, some sipunculans, and earthworms are claimed to have lost all vestiges of their larval stages (figure 16.2).

Rotifera ("wheel animalcules") is a phylum of small ciliated marine and freshwater animals. *Trochosphaera* sp. is a rotifer with a marked resemblance to trochophore larvae. We reject the idea that octopuses, earthworms, and certain sipunculans lost their larvae. We think they never had any. Rather, clams, snails, polychaete worms, and some sipun-culans acquired trochophore larvae by hybridization with rotifers. No octopuses, squids, earthworms, or other sipunculans ever hybridized.

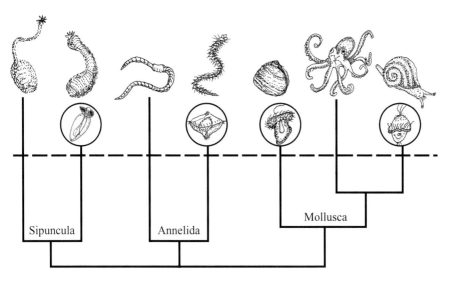

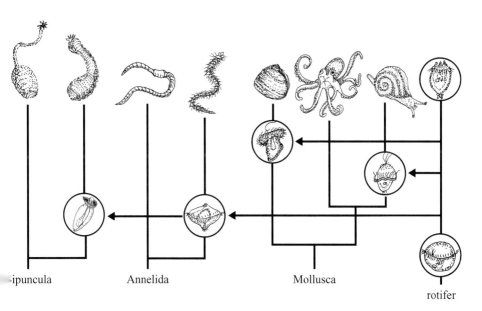

Figure 16.2
(A) Ancestors with trochophore larvae. Trochophore larvae are found in many phyla. The usual explanation is that these evolved from a common ancestor with a trochophore larva, but some lost their larva. (B) Larvae transfer explains the strong resemblance of rotifers to a trochophore larva. The rotifers hybridized with a snail and a clam, but not with an octopus. Rotifers also hybridized with a polychaete worm, but not with an earthworm. The polychaete worm then hybridized with some but not all sipunculans.

The explanation touted under the "same ancestral stock" theory is that the adult rotifers are "persistent larvae"—descendants of forms that matured in the larval state. Were this so, one would expect the genes of the lost adult to be preserved as "junk DNA" found in many genomes, but in fact rotifers have remarkably few junk genes. Why do larval forms of three different phyla resemble a rotifer in a separate phylum more than they resemble the adults into which they develop?

Echinoderms

Echinodermata are members of the phylum of animals that have radial symmetry, including starfish, brittle stars, feather stars, sea urchins, and sea cucumbers (figure 16.3). Not only are echinoderm larvae, in contrast to adults, bilaterally symmetrical; some of them resemble the tornaria larvae found in the phylum Hemichordata (acorn worms). The widely accepted explanation of this anomaly, put forward by Ernst Haeckel (1866), states that bilateral larva of echinoderms and hemichordates evolved from a common ancestor with tornaria larvae. An ancestor of modern echinoderms evolved radial symmetry, Haeckel claimed, in response to sedentary life. Fossils and larvae described since Haeckel do not support his views on the origins of echinoderms. We now know that pentaradial echinoderm adults abound in the marine fossil record for at least 540 million years. Some hemichordate larvae resemble

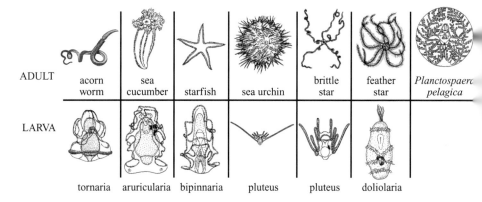

Figure 16.3
Similar larvae develop into very different adults (sea urchin and brittle star), and some very different larvae develop into somewhat similar adults (brittle star and feather star). We posit that such anomalies are not explicable by the assumption of random mutation causing "descent with modification" from a common ancestor followed by "natural selection."

trochophores, not tornarias. The larva of the echinoderm sea cucumber resembles a tornaria more than it resembles other echinoderm larvae. Adult sea cucumbers show a mixture of bilateral and radial features. A Haeckelian interpretation suggests that sea cucumbers are the living echinoderms nearest to the ancestral form. Paleontology, however, tells us that sea cucumbers evolved comparatively late in echinoderm phylogeny. We propose that larvae were later additions to the phylogeny of the adult taxa. The adult from which echinoderm larvae evolved is *Planctosphaera pelagica*, the only known planctosphere. Planktonic and spherical (as its name suggests), and up to 25 millimeters in diameter, it is propelled by convoluted bands of cilia. This animal is classified as a hemichordate because of its resemblance to a planctosphaera-like tornaria larva, but we regard it as an adult member of the group that gave rise to tornaria larvae by hybrid transfer. We claim that an ancestor of *Planctosphaera* hybridized with an acorn worm to produce an acorn worm with a tornaria larva. This type of larva was then spread by cross-fertilization between an acorn worm and a sea cucumber. Further hybridizations occurred between a sea cucumber and a starfish, between a starfish and a sea urchin, and between a sea urchin and a brittle star. The doliolaria larva of feather stars cannot be traced back to a planctosphere. The evolutionary dispersal of planctosphaera-like tornaria larvae is independent of adult echinoderm evolution.

Molecular biological evidence for larval transfer is now appearing. The sequence tornaria (acorn worms)–auricularia (sea cucumbers)–bipinnaria (starfish)–pluteus (sea urchins and brittle stars) was recognized by MacBride (1914) and appears in many textbooks. This same sequence has now appeared in the distribution of a ribosomal gene, 18S rRNA (figure 16.4). In the middle of the cladogram is the sequence acorn worm–sea cucumber–starfish–sea urchin. This is the same sequence as shown by the larvae of these groups, but it bears no relation to the evolution of the adults. Our explanation is that the 18S ribosomal gene has been transferred between taxa often, and, in the case of hemichordates and echinoderms, it seems to have been transferred at the same time as genes specifying larval form. Hybridization provides a plausible method for the simultaneous transfer of these ribosomal and nuclear genes. The larval-transfer hypothesis that was founded on animal morphology is also compatible with molecular evidence.

More anomalies in echinoderm development are depicted in figure 16.5A. Brittle stars and sea urchins are very different as adults but share a unique form of larva, the pluteus, with slender arms supported by

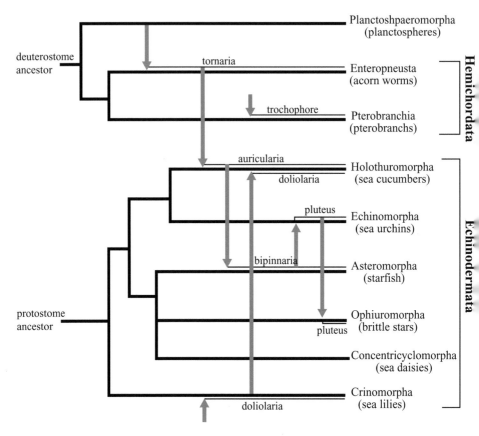

Figure 16.4
Molecular evidence supports morphological evidence. This cladogram of 18S rRNA traces lineages and estimates divergences. Cladograms often place animals in unexpected relationship patterns. This cladogram strongly supports the likelihood that genes were transferred across large taxonomic distances. The hypothesis of larval transfer is supported far better by this cladogram than is the concept of "descent with modification." (See arrows.)

calcareous rods. Our hypothesis of larval transfer explains the great anomaly that similar larvae develop into dissimilar adults. The basic pluteus larva evolved only once, in a sea urchin, and as the descendants of this sea urchin evolved, so did the larval form. An ancestor of most existing brittle stars then acquired a pluteus larva by hybridizing with a sea urchin. However, Kirk's brittle star develops directly from the fertilized egg and entirely lacks a bilateral larva. It seems that none of its ancestors hybridized with a sea urchin. Claiming that very different larvae evolved into nearly identical adults is, to us, like claiming that the

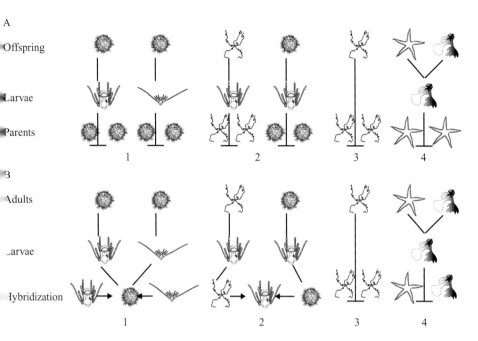

Figure 16.5
Echinoderms and their larvae are best explained by hybridization. (A) Inconsistencies in life history of echinoderms: (1) Similar adults develop into different larvae. (2) Different adults form from similar larvae. (3) Some echinoderms (e.g., brittle star) lack a larval form. (4) In *Luidia sarsi* the adult hatches and the larva remains viable for months. More than 150 years of attempts to base taxonomic classification on larvae and their adults as a single unit have not succeeded. (B) How these inconsistencies are resolved by hybridization: (1) One genus of sea urchin, *Lytechinus*, hybridized with two different species, close relatives developed very different larvae. (2) The brittle star and the sea urchin each hybridized with the same organism, giving these two unlike adults the same larva. (3) Kirk's brittle star never hybridized. It develops radial symmetry directly from the egg. (4) The overlapping metamorphosis of *Luidia sarsi* is explained by an earlier hybridization of two organisms that merged genomes.

Mandarin Chinese and Japanese languages independently evolved into Spanish.

Two sea urchins, *Lytechinus variegatus* and *Lytechinus verruculatus*, are so similar that they are classified in the same genus. But these similar adults develop from very different pluteus larvae. That very different larvae evolved independently into nearly identical adults contradicts evolutionary experience and cannot be tested. However, our suggestion that the two very different larvae were acquired comparatively recently by hybridization with the two very similar adults, as shown in figure 16.5B, is eminently testable.

Perhaps the strangest assumption of neo-Darwinists is that the five-armed tube-footed starfish *Luidia sarsi* (plate XI), which decorates the cover of *The Origins of Larvae* (Williamson 2003), evolved directly from a small, bilaterally symmetrical, translucent larva. As in other starfish, *Luidia*'s fertilized egg develops into a bilateral larva that contains a small five-armed starfish. The juvenile starfish then migrates to the outside of its larva. Whereas in most starfish the larva settles and degenerates, and the juvenile crawls away, in *L. sarsi* the juvenile drops off from its swimming larva. Five-pointed starfish and bilaterally symmetrical translucent larvae continue to live independently for months. These two very unlike organisms are the same individual, hatched from the same fertilized egg! How can a single individual split into dissimilar but recognizable animals if both have exactly the same genome? Isn't the idea that animals with larvae are descended from hybrids that have fused larval-adult genomes much more tenable? The coexistent swimming larva and crawling starfish of *L. sarsi* evolved by the merger of genomes of two animals that hybridized to produce one animal with a larva. The two genomes, however, retained a considerable degree of independence. Both genomes are expressed together, but sequentially, during development. Such overlapping metamorphosis as we see in *Luidia sarsi* occurs in other echinoderms and in polychaete worms, but it does not last as long. In our view, larval transfer explains these developmental phenomena far better than the assumption that disparate larvae and adults evolved from the same individual common ancestor.

Hypothetical Terrestrial Examples

If unfamiliar development patterns of marine animals are difficult to understand, it might help to consider some hypothetical mammalian analogues. What if kittens developed with equal frequency into dogs or cats? What if piglets or calves regularly metamorphosed into Arabian horses? What if Arabian horses developed from piglets, while Morgan horses were born as pony-like miniatures? Wouldn't we hypothesize that the immature stage is separate from the adult? In our first fictional example, dogs and cats acquired their kitten stage separately. In our second example, some Arabian horses acquired the piglet stage and others acquired the calf stage, whereas Morgan horses never acquired any immature stage.

So many adult marine animals differ so radically from the larvae from which the develop that fusion of genomes by hybridogenesis is a far more parsimonious explanation than convergence by independent random mutation. That the adults acquired larval stages from fertilizations between very different animals is a plausible, testable scenario worth of empirical refutation or verification.

Other Views on Larval Evolution

Most of the literature on evolution of larvae assumes that an animal lineage began without a larval stage, then, through evolution, developed one as a means of survival and propagation. The nematode *Caenorhabditis elegans* is capable of producing a slightly different body plan or phenotype if stressed during development. It will revert to the normal body plan when the stress is removed. Genes for this divergent body plan have been located, and it is hypothesized that continued environmental changes might cause the two body plans to diverge so much that one eventually becomes the larva that will later develop into the adult body plan. Any animal would develop directly from its egg into a young form that displayed the same body plan as the adult. Then, as conditions mandate, the development would become more and more elaborate, and indirect development would necessitate a form of metamorphosis. This explanation appeals to evolutionists who assume evolution by gradual change.

Another idea applies to particularly small animals. Large animals produce many eggs and provide them with nutrients for development. Small animals must produce fewer eggs with nourishment, produce more nutrient-poor eggs, or invent many nutrient-poor eggs that can eat. Eggs that develop quickly into life forms that ingest food are larvae. But a need (e.g., for small feeding stages) does not explain how evolution supplies that need.

The usual explanation for the similarity of the larvae of such different phyla as Mollusca and Annelida is that of convergent evolution. Just as dolphins resemble fish because they face the same environmental pressures, the similar larvae of diverse groups are claimed to resemble one another because they face the same survival tasks. One still wonders, though, how *Luidia sarsi*, the little starfish that separates from its larva and both continue to live, fits this widely accepted paradigm of the single organism that evolved a larva. But *Luidia sarsi* proclaims its double parentage for all to see.

To differentiate between common single ancestry and larval transfer by hybridization requires a comparison of DNA macromolecular sequences of both larvae and adults. Similar larvae from different adults would be expected to have similar genes that they received from their common ancestor during hybridization. Closely related adults that have different larvae should be expected to show a disparity between the genes of the two larvae, if indeed they were derived from two different sources. The picture is, of course, complicated by the formation of serial chimeras, for we feel that many present-day animals came about through not just one hybridization but perhaps many, as in echinoderms. In any case, if the tree of life is not a tree but rather a tangled web of interrelationships, the DNA evidence is expected to be confusing, like the 18S rRNA data in figure 16.4.

Consequences of Transfer

If larvae were later additions to life histories, the earliest animals lacked larvae. When a successful hybridization occurred, the resulting chimera received a coherent set of genes that each animal lineage acquired through years of natural selection. An early feeding locomotive stage in development became coupled to a larger reproductive stage. The potential to produce many more offspring of its kind resulted in the expansion of diversity known today to characterize animals with larvae. Occasional successful hybridizations between adults occurred in the seas, where eggs and sperm were distributed abundantly. We don't believe that this activity was limited to the distant past. Brittle stars in the same genus that have different larval forms indicate recent hybridization. Even close relatives may not have all their ancestors in common, if we are correct. The web (not tree) of life has both fused branches that involve many phyla and much more recent fusions that separate species within a genus.

We can now answer the question, posed by Sly et al. (2003), "Who came first, larvae or adults?" We are convinced that larvae were later additions to life histories, such that adult mollusks and echinoderms developed from eggs in the Paleozoic era before either had larvae. The basic features of larvae, however, are thought to have evolved long before animals with larvae existed (Williamson 2003), just as "the functions now performed by cell organelles are thought to have evolved long before eukaryotic cells existed" (Margulis 1993). Adult rotifers preceded mollusks, and adult annelids acquired trochophore larvae. Neither rotifers nor planctospheres nor onychophorans are persistent larvae; rather, as adults they have supplied the sources for a diversity of marine larvae.

Errant sperm do fertilize eggs from very different animals. In two of our experiments, eggs of an ascidian (a member of the phylum Chordata and therefore related to humans) were fertilized with the sperm of a sea urchin, a member of the echinoderm phylum (Williamson 2003). In another of our experiments, eggs of a sea urchin also were fertilized with ascidian sperm. In the first experiment, many fertilized ascidian eggs developed as sea urchin larvae, only three of which grew into fertile adult sea urchins. The great majority of the sea urchin larvae, however, retracted their arms to become spheroids, which failed to develop further. In the second experiment, all the fertilized sea urchin eggs began to develop as sea urchin larvae but then reabsorbed their arms to become spheroids. A few of these developed to resemble juvenile ascidians. They developed anterior adhesive organs, the first step in metamorphosis to a sessile life style in the ascidians. But they did not develop further into adult ascidians. Genes that code for larval and adult forms remain capable of expression after hybridization. This observation is consistent with larval transfer but not with the same-stock common-ancestor assumption. The fact that these hybridizations occurred in the laboratory leads us to conclude that they also took place in nature. We are optimistic that continued investigation will prove the validity of our "evolution by hybridogenesis" hypothesis; eggs of dramatically metamorphosing animals harbor at least two once-separate (heterologous) genomes that are expressed in sequence.

17

Origins of the Immune System

Margaret J. McFall-Ngai

Evolution in crowded environments leads to surprising developments. Here McFall-Ngai argues for the striking idea that even the vertebrate immune system may be an organ of symbiosis.

Over the eons, animals, plants, and all other multicellular eukaryotes diversified within the context of a bacteria-rich environment. (See plate XII.) Evolutionary selection pressure was exerted by prokaryotic and other microbes on larger organisms, especially animals. These organisms had to sense and respond to biotic cues and stimuli in marine settings steeped with life. In animals, a complicated reaction to the biotic world involves the activation of a multifaceted immune system. The vertebrate immune system evolved in the presence of bacteria that were sensed, responded to, and integrated into it. As most animals are digestive tubes—from mouths through stomachs and intestines to anuses, covered by integument such as skin—the animal immune system probably evolved, in my view, from the exigency of distinguishing healthy necessary bacterial inhabitants of these tubes and coverings relative to opportunistic, potentially dangerous hungry exploiters. Although animals possess both types of coevolved partners—those in symbiotrophic symbioses and those that are potential threats in necrotrophic associations—the vertebrate immune system mainly evolved from bacterial interactions that nurture or tend to destroy physiological homeostasis (often syntrophies) of co-evolved animal-microbial partnerships. Our concepts of animal biology change rapidly when we recognize that the animal kingdom comprises far more than single types of eukaryotic cells. More accurately, it comprises the interconnected, coevolving sets of communities with one principal eukaryotic cell type and often more than several hundred distinct kinds of bacteria. The findings that the integumentary (skin) and the digestive systems of the average human body harbor a coevolved consortium of 10^{11} skin bacteria and 10^{14} digestive bacteria demand that biologists reconsider their most basic concepts of the physiology of these systems.

Perhaps the most obvious and dramatic changes will be seen in our view of the form and function of the immune system, as a principal response to the biotic environment. The growing appreciation of the true nature of animals renders obsolete the traditional idea of the immune system as a "non-self" recognition system. Most notably, humans require the partnerships of many different kinds of bacteria (phylotypes) for normal growth, development, and homeostasis. Our understanding of the immune system must be expanded to include these new findings. Let us step back and reconsider the function of the immune system in light of related basic biological principles.

What similarities and differences exist in the ways animals respond to physical, chemical, and biological environmental stimuli? The primary purview of the animal nervous system is to modulate activity in response to abiotic, or physical changes. Three major types of neurons carry out this task: sensory neurons, interneurons, and motor neurons. Sensory neurons act as transducers of the environmental information (e.g., changes in light, changes in temperature, mechanical stimuli) and convert it into a usable form. The interneurons integrate the input so that an eventual appropriate response is made. The response or output is the job of the motor neurons, which mediate the activity of the effectors (e.g., muscles, chromatophores, bioluminescent organs). Of course, much of the information carried by these abiotic forces provides the animal with knowledge about the state of the biotic world, such as the presence of predators or prey, competitors, and food.

Animals use light to respond to the biotic world (e.g., predators, prey, mates, young, fruits, seeds). In the big picture, perhaps light is secondary for all five kingdoms, but it is crucial to invertebrate and vertebrate animals as a response to other living creatures and their products released into the environment. The same might be said for temperature among communal vertebrates. The immune system seems to respond to antigens, to biotic products of metabolism, and to prokaryotes.

The immune systems of animals also have sensors, integrators, and effectors that respond to interactions with citizens of the microbial world. The emphasis of study has been on microbial pathogenesis (necrotrophy). The most widely held belief is that the immune system views microbes as germs sensed by certain animal cells. The information would be integrated in such a way that the entire animal responds to rid its body of the encroaching dangerous microbe. This viewpoint changes with our increasing awareness of the importance of our normal microbiota in health. The normal microbiota of an animal is an integral, coevolved portion of the whole. The microbial component is an essential part of the developmental response. For example, as the mammalian gut encounters bacterial, chemical, and other antigens ingested with food, the effector system that responds to the foreign biotic world involves a tripartite dialog among its epithelial cells, its immune cells, and the normal microbiota of the gut. The response is not confined to those portions of the immune system that carry out surveillance of the gut.

Comparing Animal Immune Systems

As new ideas of microbial interactions with animal immune systems take shape, comparative biology sheds light both on aspects common to most animals and on how animals differ in immune function—that is, on how the diversity of the system evolves. The most obvious difference in immune systems across the animal kingdom is the restriction of the combinatorial (adaptive) immune system to the gnathostome vertebrates. The combinatorial immune system correlates with the presence of some specialized genes (abbreviated RAG) at the agnathan-gnathostome transition (the transition from the jawless lampreys and hagfish agnathans to the vertebrates with jaws). Macromolecular sequence analyses of both DNA and proteins suggest that this evolutionary process involved lateral gene transfer from bacteria. With the integration and natural selection into the animal genome of the RAG genes, the vertebrate immune system became capable of V(D)J (variable-disjoined) recombination—that is, rearrangement of the immunoglobin proteins to generate infinite diversity of the antibody repertoire. Immunologists assume that the combinatorial immune system is an "advance" in evolution of the immune system that renders gnathostome vertebrates "stronger"—that is, capable of more sophisticated non-self recognition. If we integrate concepts of the interaction of coevolved microbiota, does this anthropocentric viewpoint continue to be valid? Two issues that bear on this question are the position of the non-gnathostome-vertebrate animals within the animal kingdom and their patterns of the interaction with microbiota.

More than 96 percent of the diversity of the animal kingdom at both the phylum level and the species level lies with the invertebrates. However, perhaps because many biologists deal principally with a few invertebrate "models," such as *Drosophila melanogaster* (fruit fly) and

Caenorhabditis elegans (nematode worm), the major theory as to how the invertebrates survive to reproduce without the combinatorial immune system is that they are "r" selected. A typically "r"-selected animal is small and short-lived and has many young, whereas "K"-selected animals tend to be large and long-lived and to have few young. Knowledge of invertebrate diversity reveals that this assumption is naive. Invertebrates have representatives with every known life-history strategy.

How do invertebrates as abundant as the gnathostome vertebrates rely entirely on the innate immune system to interface with the microbial world? Perhaps there is a difference—reflected in the form and function of the immune system—in the ways that invertebrates and gnathostome vertebrates have evolved to interact with microbes. Let me briefly analyze the trends in animal-microbe relationships in the two groups.

An obvious difference between invertebrates and vertebrates lies in the propensity to form intracellular associations with microbes. To my knowledge, no reports confirm the presence of intracellular bacteria as components of the normal microbiota of vertebrates, although several pathogens adopt this lifestyle (e.g., *Treponema pallidum*, *Listeria monocytogenes*, *Toxoplasma gondii*, *Mycobacterium tuberculosis*). By contrast, intracellular bacteria occur as components of the coevolved microbiota in members of the major invertebrate phyla. Approximately 11 percent of all insect species have intracellular bacteria associated with the fat body (their liver analog) that provide essential nutrients to the animal.

Invertebrates, but not vertebrates, have a propensity to form binary associations with bacteria, i.e., associations in which a monoculture of specific bacterial symbionts occurs in a restricted portion of the body (often a symbioorgan), either intracellularly or extracellularly. The vast majority of the intracellular relationships mentioned above occur as binary associations. Extracellular heterotrophic binary associations have been reported in eight of the ten major invertebrate phyla. (Both autotrophic and heterotrophic intracellular and extracellular associations have been studied in coelenterates (e.g., *Hydra vividis*) and *Porifera*, green sponges.) Extracellular binary associations are rare in the vertebrates, by contrast. Independently evolved symbioses of different kinds of luminous bacteria with only distantly related teleost fishes are notable exceptions.

Invertebrates appear to have a proclivity to form intracellular or extracellular binary associations with bacteria, whereas the gnathostome vertebrates (e.g., fish, amphibians, reptiles, and mammals—backboned animals with jaws and teeth) have complex coevolved consortia that live along the apical surfaces of the mucosal epithelia and are associated with at least eight of their ten major organ systems. The microbiota appears to persist even under extreme conditions such as starvation. Although no entirely representative survey of nearly forty invertebrate phyla is available, consortia have been reported in arthropods, in mollusks, in echinoderms, in coelenterates, and in sponges. In the insects, where gut consortia have been studied extensively under experimental conditions, it appears that most elements of the consortia are "tourists"—i.e., that the microbes take advantage of nutrient-rich environments, such as the gut. When the insects are starved, much of the community is lost, and the community's composition is drastically altered by a change in diet (Broderick et al. 2004). Some invertebrate species appear to lack coevolved microbiota (Boyle and Mitchell 1978; Garland et al. 1982). Exceptions to these trends include xylophagous roaches and termites, which harbor complex coevolved microbial communities in their hypertrophied hindguts. However, the extracellular community of microbes is kept at a distance from the animal tissue by the chitinous layer that lines the digestive system. The community, including both bacteria and amitochondriate protists, is shed with each molt of the insect. Consortia in vertebrates, such as those in the mouth, occur in intimate association with the epithelium and with immune cells. They are not periodically shed.

There may be systematic differences in the ways invertebrates and vertebrates tend to interact with the microbial world. Binary intracellular or extracellular relationships are common in the invertebrates, whereas stable, complex coevolved consortia are rare; the opposite may be true of vertebrates. Far more investigation is needed.

Are these differences between invertebrate and vertebrate animals reflected in the nature of their immune systems? What is the relationship between the normal microbiota of animals and their immune systems? How do animals maintain their alliances? Invertebrates rely on an innate immune system that recognizes common molecular features of microbial cells: membrane lipopolysaccharides and cell-wall peptidoglycan, molecules unique to bacterial surfaces. Once such molecules are detected, the innate immune system clears the bacteria from the invertebrate's body. Invertebrates may circumvent activities of the innate immune system by forming alliances that promote specific associations by intracellular incorporation of one or a few types inside the body's protected sites, such as the interior light organ of squids, beyond detection of the immune system. Special, limited associations with only a few bacteria have evolved, for example in weevils with the *Sitophilus* primary endosymbiont (SOPE) and in wasps with *Wolbachia*.

How, then, do vertebrates maintain complex coevolved consortia? Such maintenance may be controlled by the adaptive immune system or by the interplay of the innate and adaptive immune systems. I suggest that selection pressure on the evolution of the adaptive immune system functioned to mediate animal-microbial interactions to maintain them in healthy balance. Although we can never reconstruct the history precisely, certain features of extant vertebrates should be detectable if I am correct. If an adaptive immune system is a shared, derived character of vertebrates, then complex coevolved microbial consortia should also be a shared, derived character. Evidence of a tripartite dialog among the microbiota, the immune system, and the animal is accumulating. In the past ten years, molecular techniques have made it possible to identify and estimate the number of resident microbes and to detect their metabolic activity. We now know that coevolved microbial consortia that associate with mammals (and perhaps with other vertebrates) may have thousands of members. Vertebrates' immune systems devote more than half of their cells to interactions with epithelia mucosa that support these consortia. Maintaining the critical interface between the innate immune system and the combinatorial immune system requires incessant interaction with normal microbiota. The microbial gastrointestinal consortia are not "commensal" for the vertebrate; that is, they do not merely "eat at the same table" as the animal that harbors them, with little effect on that animal's health and reproduction. There is strong and extensive evidence that growth and metabolism of specific microbial associates are essential for the development of animal tissues in which they reside. Microbes, far from being "enemy germs," are critical for the development of healthy immune systems in vertebrates.

Surveys of the animal kingdom may support or refute my idea that vertebrates, from their origin, coevolved with specific microbial consortia less frequently than invertebrates. The ratio of required "residents" relative to facultative, transient tourists should be estimable in any given consortium. To what extent is the composition of the consortium a direct response to the chemistry of the animal body alone, relative to the biochemistry of the tissue under the influence of the coevolved microbial alliance? I predict that the innate immune system in general will be distinguishable in vertebrates relative to invertebrates—i.e., that invertebrates are less likely to develop immunological "tolerance" that welcomes specific "others" into their bodies and does not indiscriminately kill all foreigners.

I suggest that invertebrates and gnathostome vertebrates differ in the ways they sense and interact with the ubiquitous microbial world in which they are embedded. The immune systems of these two groups are distinguishable: Invertebrates limit their interactions with microbes to only a few species; i.e., their approach is "restrictive": to rid themselves of all that dare enter. By contrast, an individual vertebrate associates with several microbial communities, each composed of hundreds of distinct individually recognized microbial groups (phylotypes). The vertebrate approach is "permissive." If my idea is valid, the primary selection pressure on invertebrate immune systems results in recognition of non-self, whereas the primary selection pressure on vertebrate immune systems involves subtle control of the complex set of microbial communities required by vertebrate animals for survival.

18

Medical Symbiotics

Jessica Hope Whiteside and Dorion Sagan

The medical sciences are enhanced by incorporating new knowledge from evolutionary biology. Here Whiteside and Sagan argue that modern medicine should reflect the new understanding of humans as multigenomic beings with an ancient evolutionary history.

Because of the difficulty of isolating variables in a human being, one of the most complex systems in the universe, medicine remains an art as well as a science. The introduction of complementary and alternative medicine into traditional Western medicine, derided by some, has the potential to broaden and make more subtle an imperfect Western tradition whose history includes medieval barber surgeons and the misrecognition of mercury poisoning as a major symptom of the syphilis that mercury was administered to treat (Fleck 1935).

Western medicine tends to favor a largely unexamined metaphor of illness as a battle or war with the fortress under siege as the sick individual; curing, here, means ridding the body of its "invaders" (Sontag 2001). The basic framework for such a battle, although given credibility by Pasteur's discovery of pathogenic microbes, was intact long before the discovery of actual invasive agents. In the Middle Ages, demons, not bacteria, viruses, or protists, were considered to be the culprits; sickness, especially during the Black Plague, was allied with evil forces and with punishment. The medieval idea of an assailed body, attacked by demons, was given substance by the discovery of microbes, but the underlying metaphor—that of an integral system attacked and brought down from the outside—is questionable at best. Ten percent of our dry weight is bacterial, vitamin B_{12} is synthesized by bacteria, and, increasingly, mitochondria (former bacteria) are recognized as of essential importance to our metabolism. Mitochondria, always present and active in mammalian cells, are related to our longevity and our optimized immune system.

Human health is predicated not on purity or on lack of contamination of one species by others, but on the many organisms that make up a healthy human body behaving together with energetic efficiency as a community.

Open Systems

The animals studied by zoology and medical science are not islands; like all complex life forms, they are open thermodynamic systems, crucially dependent on the timely intake and output of gases, liquids, and solids. In addition, social animals exchange signs, ranging from chemical phero-mones (in social insects) to leaking hormones (e.g., androgens within the social hierarchies of the naked mole rats) to symbols (e.g., the ink-marked arboreal by-product of this page). Organisms may be considered entropy-producing thermodynamic systems whose complexity correlates with their efficiency in producing molecular chaos (i.e., entropy). Experiments have shown that not only thermoregulating mammals but also the eggs of fish (loach), when stressed, undergo maximal but not optimal entropy production (Zotin 1972). The spontaneous organization of non-living complex thermodynamic systems, a process which is observed to coincide with their becoming more effective entropy producers, suggests that reducing a gradient is a natural function of organisms and of nonliv-ing complex systems (Sagan and Whiteside 2004). Ecosystems also show that complexity, measured by species richness and extent, correlates with thermoregulatory prowess. The most complex ecosystems, rainforests, may be interpreted as transpiration machines that keep themselves cool in the face of external heat, a process that produces more entropy than would be the case in the absence of such complex systems. According to the second law of thermodynamics, the "normal state" of things, includ-ing the universe as a whole, is in equilibrium. Such stasis, total random molecular chaos, might be considered "health" from a physical point of view. The organization of organisms does not contradict the ubiquitous entropic process, but aids them by contribution of wastes (which can then be recycled to continue the flux of material and energy) to the sur-roundings. Organisms are "exergy machines" (Minkel 2002); the back-drop of living organization is not just genetic, but also thermodynamic (Wicken 1987; Schneider and Sagan 2005).

While growth, health, intelligence, and asymptotic approach to a species-specific "design" are normal, so too are disease, death, and decay. These latter wasting processes are to be expected in a universe where the second law of thermodynamics is as important as gravity, and arguably more important at the mesoscale inhabited by live organisms. Interpreted naturally, life persists in a world tending toward molecular chaos not because it is a divine aberration, but because the energetics of complex systems producing more entropy than would occur in their absence is favored. Moreover, because organisms are open systems, they are more likely to form alliances so as to pool metabolites, proteins, and genes. When and where such alliances can persist, producing more entropy than would otherwise be the case, they will tend to proliferate relative to their unallied comrades, in part because of their superior access to the energy and material that provides them sustenance upon which they maintain and grow.

Impure Origins

In the late nineteenth century, Nietzsche complained in *The Genealogy of Morals* that the meaning of the word "good" underwent a moral change in the wake of Christianity: whereas, originally, it had meant "strong," or "noble," and was contrasted with "bad," it now was loaded with values such as charity and chaste love and was contrasted with "evil." In short, according to Nietzsche, what was good and what was evil were switched. The situation regarding health and sickness is very similar. When Kwang Jeon received a new batch of amoebae from his colleagues, he noticed that one of his collections was sick—that the amoeba population had been invaded by bacteria. After many generations of amoeba growth and experimentation, Jeon found that the nuclei of his newly afflicted amoebae had become dependent on their intracellular bacterial invaders. The nuclei, now accustomed to their invaders, would expire in a few days. However, they could be rescued by injection into the cytoplasm of bacteria. A "bacterized" amoeba had evolved from a normal one free of bacteria. Here we have an example of an infection or a disease microbe not just tolerated but, with time, required for amoeba survival. Once dismissed as fantasy, it is now widely accepted—partly on the basis of comparisons of gene sequences and partly on the basis of morphological, chemical, and behavioral similarities—that all eukaryotic cells arose from similar "infections": infections of anaerobic archaebacteria, first by motile eubacteria (probably spirochetes; see chapter 13) and then by the oxygen-breathing ancestors of mitochondria. Oxygen-respiring bacteria—thought to have proliferated with the accumulation of free oxygen, the new waste product of water-using cyanobacteria, in the atmosphere—would have provided the anaerobes they infected with a paradoxically life-giving trait: the ability to detoxify oxygen, a potentially fatal poison to anaerobes. In short, the bacterial invasion of organisms by other organisms does not lead necessarily to death but can, if the infected organisms survive, lead to new phenotypic and genotypic traits. From a thermodynamic perspective, the destruction of microbes by invaders is a viable entropy-producing strategy only if the supply of the relevant microbes is virtually inexhaustible. Like a pioneer species growing rapidly to lay the groundwork for an ecosystem, the invader may grow fast in the short run. But, as with pioneer species, for invaders or infectious agents to survive over the long term they must not destroy the other organisms on which they depend for survival. Because negative signs of pain and illness have been central to medicine's awareness of malfunction, non-pathological community relations are generally ignored in the practice of medicine. Less effective emergency-room treatment is favored over preventive medicine.

All familiar macroscopic organisms are superorganisms. In view of the entropic function of living systems, descriptions of cows as roving methane tanks and of other animals as mobile energy extractors must be taken seriously. Digestion depends on a community of intestinal bacteria, and much of our dry weight is bacterial. (See chapter 17.) The superorganismic nature of the body was made clear in a thought experiment by the biologist Clair Folsome. Folsome asked us to imagine a device that would cause our animal cells to disappear instantaneously. He then asked:

What would remain would be a ghostly image, the skin outlined by a shimmer of bacteria, fungi, roundworms, pinworms and various other microbial inhabitants. The gut would appear as a densely packed tube of anaerobic and aerobic bacteria, yeasts, and other microorganisms. Could one look in more detail, viruses of hundreds of kinds would be apparent throughout all tissues. We are far from unique. Any animal or plant would prove to be a similar seething zoo of microbes.

A human being is a cohesive mass of millions of cells regulating at 37.5°C (98.6°F). Bacteria, larger microbes, social insects, and even rodents construct living units of members that exist only in the aggregate. In serial endosymbiosis theory (Margulis 2004), we trace the cells that we consider our own animal cells from two or perhaps three or four separate bacterial ancestries. The oldest, presumably, is the *Thermoplasma*-like lineage—relatively large, oxygen-poisoned bacteria that are classified with archaebacteria. Tellingly, the archaebacteria (also called archaea) are so genetically distinct from other bacteria that Carl Woese considers them as different from normal bacteria as bacteria are from animal-like cells with nuclei, true chromosomes, and mitotic cell division. Yet these same "strangers in our midst" bear histone-like proteins and other fossil fingerprints that relate them to ancestral animal cells.

Today, mitochondria, inherited only from the mother and metabolizing oxygen in virtually every cell of the body, retain their own DNA, reproduce on their own timetable separate from the rest of the cell, and generally behave like bacteria. It is accepted in textbooks that they were once bacteria—bacteria that became symbiotic, merging strains, to form a new, more powerful living whole. More tenuous evidence, such as the elaborate intracellular motility systems in our kind of nucleated cells, suggests a third partner—one so old and integrated that its symbiosis is now almost undetectable. The complex intracellular dance of mitosis and meiosis, as well as the similarity between spirochetes in the muds and interstices of animal bodies, and the wriggling congenital appendages of cells with nuclei, may suggest that spirochetes long ago attached to, entered, and lost parts of themselves in the origin of intracellular and locomotory movement within nucleated cells. *Mixotricha paradoxa*, a cell once classified as a "ciliated protozoan" but now known to be a trichomonal protist, is actually a composite of internal symbionts, attached waving appendages, and would-be separate spirochetes, permanently attached in a manner that provides the motley assemblage with the wherewithal to get where it is going. One can imagine similar partnerships flourishing 2,000 million years ago, when oxygen was a poison, mobility was a precious commodity, and oxygen-using bacteria and fast-swimming spirochetes often made the difference between life and death. Because spirochetes are anaerobic and often symbiotic today, and are theoretically well integrated into the structure of the mitotic cell, they may be the oldest symbiosis of all. Like the Cheshire Cat fading to a smile, our oldest bacterial ancestors blended so well, perhaps, that of their original individual presence only the slightest signs of them remain.

Although it is not possible to distinguish *Mixotricha paradoxa* from *Trichonympha* sp., from *Deltatrichonympha* sp., or from other hindgut protists in the fossil termite *Mastotermes darwiniensis*, intestinal symbionts very similar to those of today are detected in Cenozoic Period millipedes (e.g., polyxenids or tropical termites, figure 18.1). Both spores and filaments that resemble those of *Arthromitus* (*Bacillus anthracis*, as Jeremy Jorgensen (1999) showed) in extant arthropods have been found

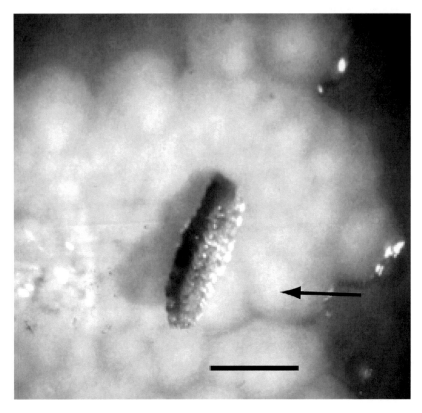

Figure 18.1
Polyxenids (arrow) fixed and stored in formaldehyde for 5 years surrounded by many light circular colonies of live intestinal bacilli. The death of the arthropod and storage in the embalming fluid did not kill or even harm the bacteria. Anthrax bacilli (e.g. *Bacillus cereus, Bacillus anthracis*) resist embalming fluids, boiling, desiccation, freezing, and many other environmental insults that are harmful or lethal to the animal. Scale bar = 1 cm.

inside the intestines of common insects and their relatives (figure 18.2). As Joseph Leidy (1850) described, intestinal symbionts have specifically associated with millipedes and insects for at least 20 million years, the age from which they have been preserved in amber. Since termites and microbes existed in the Cretaceous, these filamentous intestinal associations probably have persisted for 100 million years.

Whether or not sufficient evidence mounts to prove spirochetal ancestry of eukaryotic cells (chapter 13), there is no doubt that throughout the evolutionary saga organisms of different kinds have often come together in permanent union (Fry 2000). The thermodynamic vicissitudes of energy flow reward groups of organisms working more effectively to

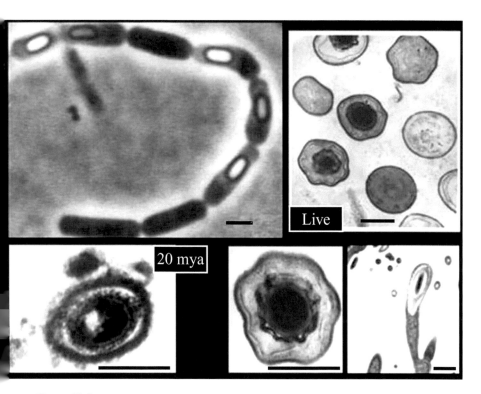

Figure 18.2
The *"Arthromitus"* (*Bacillus cereus*) from the intestine of the sow bug *Porcellio scaber*. Upper left: *Arthromitus*, the filamentous intestinal form of the anthrax bacterium *Bacillus anthrasis* and *Bacillus cereus*. Upper right: spores of the bacillus at various stages of maturity. Bottom left: a 20-million-year-old (Miocene) *Arthromitus* spore from the fossil termite *Mastotermes electrodominicus* preserved in amber (from Santo Domingo, Dominican Republic). Bottom center: a live *Arthromitus* from the Sonoran desert dry wood termite *Pterotermes occidentus*. Bottom right: an unidentified intestinal symbiont—a filamentous *Arthromitus-like* multicellular bacterium that bears spores at the tips of its branches from the hindgut of *P. occidentu*. Scale bars = 1 μm.

tap free energy and direct it, or redirect it, into their autocatalytic selves. The process applies to human groups so interdependent we cannot survive if not linked to one another or wired to machines (chapter 7). The Internet is becoming a neural net—a self-sensing, proprioceptive superorganismic organ of cyber-intelligence. "Wiring"—the connecting up of members of societies into superorganisms—has been going on for thousands of millions of years, since before the origin of eukaryotic cells.

Luxuries become necessities; experiments become organs. The electronic conduits fanning out across Earth represent a new, human chapter in the ancient evolutionary story of mergers and fusions of living beings. Individuals reduce their totipotency, and slow down to form collectives more powerful than any single member. Former confederacies become individuals at more inclusive levels of organization. There is strength in numbers, but even more in organized numbers.

An irony of this superordination of organisms sacrificing themselves to more effective group action is the medical system where financial flows from fundraising to research on curable diseases would be compromised, interrupted, or even cut off if cures, especially inexpensive cures, were to be found and administered. The "disease-industrial complex" flourishes because of individual illness. There is no absolute health and sickness, and the health of corporate medical complexes is often at odds with the best interest of the individual. Because life is composed not of discrete, impregnable monads, but of energetically open systems connecting to one another in multiple ways, identity and individuality crop up at larger and more impressively complex levels. Evolutionarily, the human individual is an organized assemblage of bacteria, remnant bacteria organized into the animal cells with nuclei (eukaryotic cells), and of these cells themselves organized into tissue, organs, and organic systems (e.g., circulatory or reproductive).

Serial endosymbiosis accounts for the origin of our animal cells, from diverse sources. But such series of symbioses, in the thermodynamic sense of complex open systems joining powers to channel energy flow, extend far beyond the realm of bacterial and nucleated cells. In retrospect, the Western notion of a self is epitomized by the loner status of a cowboy mounted on the very horse that indigenous Americans mistook as the centaur bottom of a breed of powerful new humanoid mammals. Survivors themselves are not impregnable; rather, they are open systems, requiring continuous contact with and flows in and out of the environment simply to survive. "That the self advances and confirms the myriad things is called delusion," remarks Dogen Zenji. "That the myriad things advance and confirm the self is enlightenment." (Aitken 1984)

Evolutionary Medicine

The human body is not Platonic and separate; it is ecological, microecological, and connected. Thus, in contrast to the mechanical metaphor of a machine designed for a particular purpose, the organism is a naturally evolved system with a natural function and purpose: to dissipate energy.

The 3,000-million-year-plus history of this system has given it many "ghosts": traits, no longer ideal, of biological functions matching environments no longer extant. Many of these traits are correlated with earlier, lower population levels. Humans evolved as nomadic apes, but the establishment of agriculture brought about a sedentary lifestyle that was at odds with a multi-million-year ancestry of hunting, running, physical exercise, dumping near temporary living sites, and so forth. The weed-like growth of humanity from a sparse primate to a dominant planetary mammal has led to health effects that can be understood only in terms of an evolutionary mismatch, or lag, between adaptation to previous conditions and changed recent environments. Bortz (1984, 1985) has suggested that virtually every modern ailment has a component related to lack of exercise (owing to the disappearance of a nomadic lifestyle) and to urban hyperstimulation. Examples of evolutionary lag include the fondness for sweet and salty foods, which were rare in *Homo sapiens*' ancestral environment. However, the taste for sweet and salty foods has not been mitigated to keep pace with technological innovations. The result is an increased incidence of diabetes and hypertension in urban populations. A theoretical emerging practice of medical symbiotics would explicitly recognize that health must be seen in the context of microecology (the cell communities of the body) and evolutionary change.

Positive Associations

The recognition that positive symbiosis—despite attracting little attention relative to pathogens in medical circles—is more prevalent in evolution than negative partnerships suggests that new attention should be paid to the links between symbionts and altered (perhaps even improved) physiology and behavior. Discovery of microbiota typical of obesity (the relative proportion of Bacteroidetes to Firmicutes gut bacteria being decreased in the obese) suggests opposite profiles associated with health (Ley et al. 2006). The rabid dog whose behavior is altered by the rabies virus to become mad—frothing at the mouth, biting, and helping to spread the virus—is a classical example of negative behavior. But are there opposite, positive behaviors? Increased biting means, among other things, increased exposure to genetic change through acquisition of foreign genes, proteins, hormones, and even genomes—a possible biological backdrop for the enduring vampire myth. Infectious diseases, as Moalem and Prince (2007) show, may be transmitted by positive genetic traits that are selected for and ultimately evolutionarily "worth it." As Hayden (2003) shows, James Joyce, Oscar Wilde, Vincent van Gogh, Gustave Flaubert, Abraham Lincoln, and Ludwig van Beethoven were not only afflicted with *Treponema pallidum*, the syphilis spirochete, but apparently spurred on to new heights of creativity by their infection. Theoretically, these geniuses owe some of their special traits—including an alleged mental clarity characteristic of latter stages of syphilis, whose spirochetes contain proteins similar to those found in the brain—to their disease organisms.

The candidiasis sufferer's craving for cornbread and wine, the syphilitic's lust, the cholera sufferer's explosive diarrhea, and the common cold carrier's sneeze all perpetuate disease organisms. What subtler behaviors perpetuate positive microorganismic interactions?

The cat-and-mouse relationship between microbial evolution and medical intelligence (e.g., the germ theory of disease, modern sanitation, antibiotics) is probably part of a larger story of organisms developing heightened powers of perception in response to potentially fatal interactions (Margulis and Sagan 2007). To what extent is the medical-industrial complex refractory to evidence that HIV, found without AIDS symptoms, is a viral community correlate and not the cause of AIDS? In addition, it has been suggested that AIDS, which can be triggered by malnutrition (e.g., in Africa, where testing protocols differ significantly) or by overexercise ("athletes' AIDS"), is not a general syndrome caused by a single virus, but rather reflects a shutdown of a part of the immune system that wasn't even recognized before the 1990s, when pharmaceutical and insurance companies set up profitable shop. This "first responder" immune system depends on intracellular deployment of oxygen and nitrogen compounds (Kremer 2008). Compromised mitochondria, required for immune function as well as for apoptosis, are the source of AIDS, in Kremer's view.

Doctor-caused diseases are called iatrogenic. The "magic bullet" approach of curing by identifying and exorcising biological demons via measures that may bring them on all the more dangerously in the next generation, or that may do more damage to the affected person than the invader (as was the case with AIDS "cocktails"), should be rigorously reexamined in the light of modern biology's awareness of the chimeric animal. No organism is impregnable; each functions always as part of a community of superorganisms whose behavior, physiology, and even genetics are in flux. Each fluctuates within various acceptable ranges within interacting populations of others.

Parasexual Selection

If one accepts Charles Darwin's theory of sexual selection, according to which female insects and other organisms influence evolution by choosing certain mates over others, then one must also consider the effects of choices in nonsexual and parasexual associations. Because eating is a prelude to endosymbiosis (eating, for example, is thought to have preceded the cyanobacteria symbiosis that led to algae and plants), the choice of what to eat has evolutionary consequences. Complementary medicine often mentions the value of *Lactobacillus* sp., found in the yogurt and the kefir consumed by a human population that arguably holds the record for greatest average longevity. Do populations of these milk-digesting bacteria confer symbiotic health effects? If so, what are they? Survival through the winter in the high mountains on kefir, yoghurt, and other milk products entails a supply of vitamin B in the diet. Can such advantages be replicated with other species, or improved upon, once their modes of interaction are better understood? What of the reports of turmeric (cucumin) as a powerful anti-carcinogen, or of avocado and other sulfur-containing foods as helpful in maintaining mitochondrial health? Such questions would be mainstream in the proposed field of medical symbiotics.

The biologist Margaret McFall-Ngai emphasizes the huge concentrations of microbes in the waters in which our ancestors evolved. Some of these bacteria are symbionts required by the normal digestion processes of the animals in whose guts they live. McFall-Ngai points out that immune systems may have originated from the systematic recognition of intestinal symbionts in the aqueous medium. According to this idea, the relative rarity of symbiogenesis in gnathosome vertebrates stems from the need to protect earlier specific associations in obligate ancestors.

Entropy production tends to decrease, while overall entropy production reaches a local maximum in healthy adults, in climax ecosystems and, arguably, in the evolving biosphere (Schneider and Sagan 2005). At present, humanity resembles a pathological, fast-growing agent of mass extinction that destabilizes the maximally stable ecosystem: tropical rainforest. Thus, from a global health perspective—and global health is necessary in the long run to sustain human social and individual health—humanity measurably disrupts optimal biospheric function. Global warming, an index of impaired biospheric function, lowers the biodiversity and complexity characteristic of the most efficient entropy producers. Nonetheless, illness exists along a continuum, even at the planetary level.

Phenotypic effects of association, if neutral or positive to fitness, health, and reproduction, will increase the chances of symbiosis at the permanent, genetic level. Over evolutionary time, sickness can morph into health, as attested by the infective origins of the animal cell. Clearly, a new science, "medical symbiotics," might be organized, planned, and funded that would study health from a cosmic and microecological standpoint. The first task would be to jettison the idealistic nonsense of the organism as an isolated, genetically impregnable system, a culture of one.

V
Consciousness

19

Animal Consciousness

Gerhard Roth

Sensation and complex data processing preceded human consciousness by more than 1,000 million years. Here Roth traces how thinking-like processes that began in mute cells continue through to visualizing amphibians to symbol-manipulating humans.

The question of animal consciousness is as old and difficult as the mind–body problem. While some philosophers, psychologists and even neurobiologists deny the existence of consciousness in other animals, others are convinced that some taxa of animals probably have at least some of the states of consciousness found in humans. Human consciousness includes states, ranging from simple awareness to self-reflection, that appear to be correlated with interactions of cortical and subcortical brain centers. Inside the human brain, processes that take place in the six-layered isocortex are the only ones accompanied by consciousness. Most human brain centers involved in consciousness are present in tetrapods (amphibians, reptiles, birds, mammals other than humans). Therefore, most tetrapods probably have attention, i.e., conscious awareness of sensory events. Consciousness associated with empathy or anticipation of future events may be found only in primates and cetaceans and therefore they may require a large cortex. Conscious states such as knowledge attribution, self-awareness, and use of simple syntactical language seem to be restricted to great apes and may require a large prefrontal cortex. Complex syntactical language appears to be unique to humans. Differences in cognitive functions between humans and non-human mammals appear to be related to the increased number of cortical neurons, their connections, and the speed of information processing. The evolution of the Broca speech area seems responsible for grammatical-syntactical language and for the prolonged maturation of the human brain.

States of Consciousness

Consciousness varies widely, ranging from deep coma to the highest degree of concentration and from alertness to self-reflection. The most general form is wakefulness or vigilance, usually combined with subjective awareness or conscious experience of something. This "something," the experience of one's own presence in the world, includes external stimuli, internal body stimuli (proprioception), emotions, and mental activity. Attention—increased awareness—leads to increased and focused perceptual abilities. A more special type of consciousness is body awareness, the belief that "I belong to my body." The conviction that "I am the one who existed yesterday and will persist into the future" is referred to as autobiographic consciousness. One is aware of voluntary control of one's movements and actions as the author of one's thoughts and deeds. Self-awareness is the ability of self-recognition and self-reflection.

That different aspects of consciousness can dissociate and be lost after damage to different parts of the brain points to a modular organization of consciousness. The activities of different structural and functional brain subsystems lead to different consciousness states.

The six major portions of human brains are hindbrain (medulla oblongata), cerebellum, pons ("bridge"), midbrain (mesencephalon), between-brain (diencephalon), and endbrain (telencephalon) (figure 19.1). The telencephalon of all mammals consists of a large folded cortex that covers "sub-cortical" structures. Most of the cortex, the "isocortex," has six layers. We seem to be aware only of activities in the "associative cortex," part of the isocortex (figure 19.2). But although activities of subcortical brain portions (reticular formation, the limbic system, the cerebellum, and the basal ganglia) are not accompanied by consciousness, those portions of the brain nevertheless contribute in different ways to the generation of consciousness. Neuroscience now can visualize this by means of functional magnetic resonance imaging (fMRI), a technique based on the fact that inside the brain an increase in neuronal activity (e.g., as a response to sensory or cognitive stimulation) is accompanied by an increase in oxygen consumption and local blood flow, which in turn measures the magnetic properties of the blood. The spatial resolution of fMRI is about 1 millimeter; the temporal resolution is about 1 second.

The limbic system that extends throughout most parts of the brain contains cortical as well as subcortical centers in the hindbrain (pons, medulla oblongata), in the midbrain (mesencephalon), in the

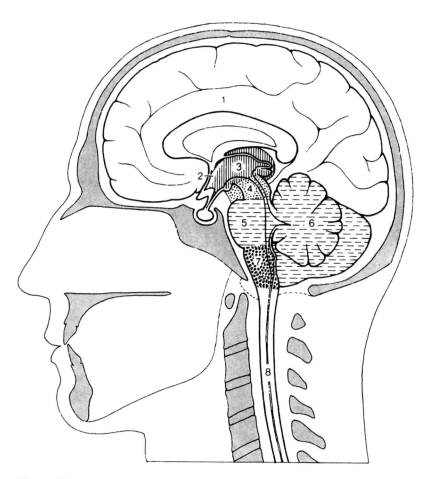

Figure 19.1
Schematic view of the major parts of the human brain and spinal cord and their positions inside the head. 1: Telencephalon (endbrain). 2: Junction between telencephalon and diencephalon. 3: Diencephalon (between-brain). 4: Mesencephalon (midbrain). 5: Pons (bridge). 6: Cerebellum. 7: Medulla oblongata (hindbrain). 8: Spinal cord.

224   Chapter 19

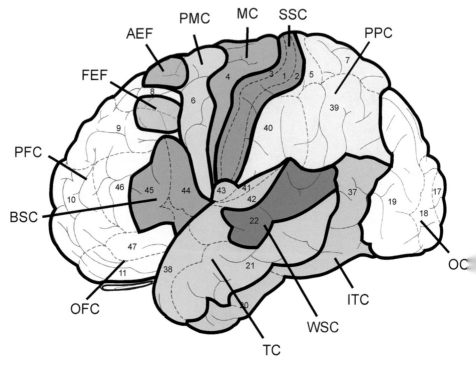

Figure 19.2
Lateral view of the human cortex. Numbers refer to organization of cells in cytoarchitectonic cortical fields. AEF: anterior eye field. BSC: Broca speech center. FEF: frontal eye field. ITC: inferotemporal cortex. MC: primary motor cortex. OC: occipital cortex. OFC: orbitofrontal cortex. PFC: prefrontal cortex. PMC: dorsolateral premotor cortex. PPC: posterior parietal cortex. SSC: somatosensory cortex. TC: temporal cortex. WSC: Wernicke speech center.

diencephalon, and in the endbrain (telencephalon) (figures 19.3 and 19.4). It is thought to evaluate what the mammal is doing, and to store the results of the evaluation in the various kinds of memory. Consciously, this is experienced as emotions and motivational states.

The amygdala, reciprocally connected with the associative isocortex, is the most important subcortical center for emotional learning of predominantly negative and threatening objects and situations and for recognition of emotional communicative stimuli (faces, gestures, etc.). The hippocampus and the surrounding parahippocampal and perirhinal cortex are centers for the formation and the consolidation of traces of the kinds of memory that, in principle, can be consciously retrieved and reported.

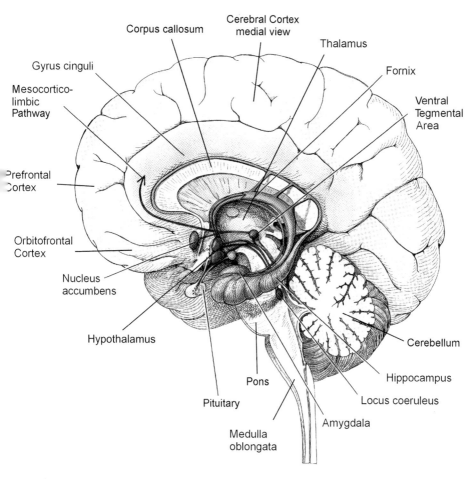

Figure 19.3
Medial view of the human brain showing major centers of the limbic system and the brain-stem reticular formation. The basal ganglia are not shown.

The dorsal parts of the basal ganglia are closely associated with the isocortex and the limbic system. They concern subconscious planning and final decision in voluntary action. The ventral parts of the basal ganglia and the mesolimbic system that are mediated by the neuromodulator dopamine with connections to the orbitofrontal cortex are involved in the formation of positive memories, pleasure, and motivation.

The anterior cingulate cortex, tightly connected with the prefrontal and orbitofrontal cortex shown in figure 19.4, is involved with the somatosensory and insular cortex in the sensation and the memory of

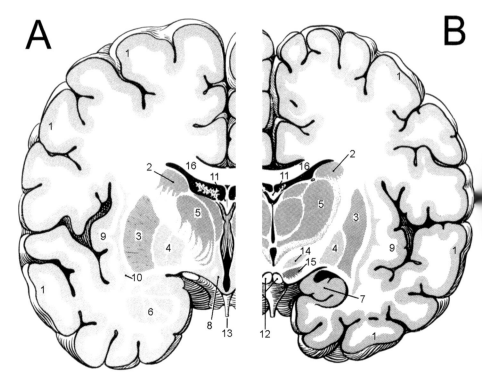

Figure 19.4
Cross-section through the human brain at the level of the corpus striatum and amygdala
(A) and at the level of the thalamus and the hippocampus (B). 1: neocortex. 2: nucleus
caudatus of striatum. 3: putamen of striatum. 4: globus pallidus. 5: thalamus. 6: amygdala.
7: hippocampus. 8: hypothalamus. 9: insular cortex. 10: claustrum. 11: fonix (fiber bundle).
12: mammillary body (part of hypothalamus). 13: infundibulum (part of hypothalamus).
14: subthalamic nucleus. 15: substantia nigra (part of basal ganglia). 16: corpus callosum.
(Modified from Kahle and Frotscher 2005.)

pain, and in guidance of attention and error monitoring. Assessment of
risks occurs here and in the orbitofrontal cortex. The left posterior parietal cortex is related to symbolic-analytic information such as mathematics, language, and meanings of drawings and symbols, whereas the right
parietal cortex correlates with spatial orientation, control of hand and
eye movements, change of perspective, and control of attention. Lesions
of the right posterior parietal cortex (mostly right hemisphere) produce
deficits in spatial orientation. Lesions in the upper temporal cortex lead
to a lack of insight or ability to interpret complex auditory stimuli,
including Wernicke's semantic speech center, which is crucial for the
understanding and the production of meaningful written and spoken

language. The inferior temporal cortex (IT) processes meaning and interpretation of complex visual information about objects and scenes. Lesions in IT lead to agnosia (left IT), color agnosia (right IT), the inability to recognize faces (right or bilateral), deficits in categorization, changes in personality and emotionality (mostly right side), and deficits in use of contextual information.

The prefrontal cortex (PFC) is involved in short-term memory, intelligence, attention, and selective control of sensory experience, strategic and associative thinking, planning and decision making, temporal coding of events, and spontaneity of behavior (Roth and Dicke 2005). Thus, the PFC is predominantly "cognitive." One important part of the (mostly left) dorsolateral PFC is Broca's speech area. The orbitofrontal PFC, in contrast, relates to social behavior, ethics, risk assessment, awareness of consequences of behavior, and emotional life (including control of behavior). The orbitofrontal PFC is oriented predominantly toward emotional and social aspects of life; its damage is correlated with loss of interest in important events; with "immorality" (e.g., indifference to sin or cruelty), and with disregard for negative consequences of behavior.

Neither the cortex *per se* nor a "highest" brain center produces consciousness. Consciousness is dispersed and is the result of large-scale interaction between many cortical and subcortical areas and centers. The associative isocortex, with its components, contributes to the diversity and content of consciousness, which is subject to strong influences from the primary and secondary sensory and motor cortices and the subcortical limbic (emotional and motivational) centers. Dreaming is a special form of consciousness. During dreaming of visual, auditory, somatosensory, or (more rarely) tactile, olfactory, or gustatory events, the respective cortical sensory and motor areas are active, as they are in remembering. It is now assumed that dreaming has directly to do with re-organization of memory contents, which are stored in these cortical areas under the influence of the hippocampus. In a similar way, introspective meditation activates specific cortical areas (e.g., the posterior parietal cortex) during the sensation of "dis-embodiment," or the prefrontal and temporal cortex during the "loss" of ego-identity.

Consciousness's Function

Conscious awareness is measured by the neural activity in associative areas. These relate to (1) sensory stimulation through certain thalamic areas, (2) adequate stimulation by other subcortical centers (e.g., the limbic system and the basal ganglia), and (3) high local metabolic activity that metabolizes oxygen and glucose from blood flow and can be seen by means of fMRI.

We cope with new, important, unexpected, and unfamiliar perceptions only if we are consciously attentive. As a consequence, certain brain regions are activated, and this neuronal activation is accompanied by increases in local blood flow and metabolism. The brain modifies the strength of synaptic coupling among neurons in the local regions. The synaptic changes lead to formation of new network properties that deal with new tasks.

Why are we conscious at all? During the evolution of primates, increases in demands of action planning, imagination, strategic thinking, and complex syntactical language probably necessitated the construction of a conscious phenomenal world (Roth 2000). The virtual actor, the ego, plans, acts, and communicates without caring about or being aware of neuronal activity. The invention of a phenomenal world probably was necessary to survival in changing environments.

Cognition and Consciousness in Vertebrates

Vertebrates appear to display focused attention and extended memory and categorization, and form cognitive maps. Among some birds and mammals at least so-called "higher cognitive functions," including concept learning, knowledge representation, analogical thinking, and abstract representation, are found (Byrne 1995; Pearce 1997). But such higher cognitive functions are not necessarily accompanied by consciousness, even in humans.

Cognitive functions that require our consciousness (and probably also that of non-humans) include the following:

(1) task-principle learning and tool learning (macaque, capuchin monkeys, apes, corvid birds)
(2) "vicariousness" or empathy (taking the perspective of other individuals in sympathy, deception or counter-deception) (e.g., monkeys, and great apes)
(3) anticipation of future events; e.g., preparation of tools in advance (great apes, perhaps some monkeys, corvid birds)
(4) comprehension of usefulness of tools (great apes)
(5) knowledge attribution/theory of mind (probably great apes, particularly chimpanzees, but this is disputed)

(6) self-recognition in mirrors (great apes, dolphins, corvid birds; Byrne 1995)

(7) distinction between appearance and reality (chimpanzees, dolphins)

(8) playing with imagined objects and teaching (chimpanzees)

(9) understanding and use of simple language (great apes, dolphins).

Complex grammatical-syntactical language seems restricted to humans.

Functions 4–9 may distinguish monkeys and great apes. Primatologists almost unanimously agree that monkeys (e.g., capuchins) use tools, but do not understand how they work or why one tool is effective and another is not. Experiments in mirror self-recognition indicate that monkeys use mirrors to look behind objects but they fail to recognize themselves. All great apes and dolphins tested show mirror self-recognition.

Monkeys and other mammals use intraspecific vocal communication that expresses relatively complicated meaning. This includes symbolic information about absent objects, prior or future events, or social relationships between mutes, enemies, or children. Sentences of at least three words are communicated and understood by chimpanzees, gorillas, and dolphins. Chimpanzees combine words to form new words, but even after training they fail to show the linguistic capabilities of 3-year-old children. Apes appear to have linguistic capabilities comparable to those of 2-to-3-year-old children. Post-toddler children rapidly develop syntactical language far superior to communicative systems known in other animals.

Brains

Are differences in cognitive abilities and consciousness due to differences in organization, size, physiology of the cortex, subcortical, or other parts of the brain? Here we compare tetrapods, including humans, in respect to brain size, organization, the isocortex, the frontal cortex, and prefrontal brain development.

1. All tetrapods, despite enormous differences in outer appearance, are very similar in overall brain organization and even in many details. All subcortical structures, physiology, and biochemistry are extremely similar. The derivation of the homology of the mammalian isocortex is not clear, but it appears to have evolved from the medial and dorsal pallium still present in extant amphibians. Reptiles and birds have developed non-layered cortical structures (called "anterior dorsal ventricular

ridge" in reptiles and "nido-, meso-, and hyperpallium" in birds), which are assumed by neurobiologists to have developed from the lateral or ventral pallium found in extant amphibians.

2. The isocortex of larger mammals, believed to be the seat of higher cognitive function, particularly in humans, has been studied extensively. There are enormous differences in absolute and relative brain size among tetrapods and mammals, whereas changes in relative isocortex size are relatively inconspicuous because cortex size increases isometrically or slightly positively allometrically with brain size (Jerison 1997). In humans, isocortex represents 76 percent of total brain mass; in chimpanzees, 72 percent.

3. Neither the frontal nor the prefrontal cortex has increased dramatically in size in humans relative to other mammals, contrary to common belief. The prefrontal cortex again exhibits an isometrical or slightly positively allometrical growth relative to total brain mass.

4. Are anatomical or physiological specializations in the cortices correlated with the human "higher" cognitive abilities? The only three features that might distinguish the human cortex/brain from other primates are differences in number of neurons and synapses, in timing patterns of growth and maturation, and in speech centers.

Human brains, relative to the brains of other primates, to have the largest number of neurons, the largest number of synaptic contacts, and, in general, the largest cortex mass. The human cortex is about four times as large as those of other great apes; it contains from 12,000 million to 15,000 million neurons and 40,000,000 million synaptic contacts. Although cetaceans (whales and dolphins) and elephants have a much greater cortical surface, their cerebral cortex is much thinner and less densely packed with neurons, and thus the inter-neuronal distances are greater. In addition, the conduction velocity of cortical axons is much slower. The combination of these facts probably accounts for the superiority of human intelligence and consciousness relative to these large-brained mammals and other primates.

Human brains require a prolonged period of development. The isocortex, with its simple afferent connections and intracortical connections, is immature at birth. Postnatal maturation in monkeys consists in an explosive increase months after birth. In humans, the rate of maturation peaks at the age of 3 years. A dramatic reduction in the number of synaptic contacts ("pruning") follows. That reduction of synaptic contacts (sometimes called "neural Darwinism") is followed by the strengthening of the neural contacts that persist. Portions of the human brains (e.g., the orbitofrontal cortex) continue to mature until the age of 20 years, much longer than in any other primate. A critical phase in human brain development occurs around the middle of the third year. Major anatomical rearrangements in the association cortex cease, and the period of "fine-wiring" (detectable by increased synthesis of "synaptic densities") begins.

Two speech centers are known: Broca's area in the frontal lobe (responsible for temporal aspects of language, including syntax) and Wernicke's area in the temporal lobe (responsible for meaning of words and sentences, i.e., semantics). It is likely that Broca's area, if not Wernicke's, is an evolutionary novelty. In all mammals, destruction of Wernicke's area in the left temporal lobe regions leads to vocal communication deficiencies; thus, the Wernicke speech (or, better, "intraspecific communication") area seems to be a common feature. In contrast, at least the anterior part of Broca's area appears to be exclusively human, while the posterior part of the Broca speech center in humans and the ventral premotor area of non-human primates, with similarities in tissue and cell architecture, are probably homologous. This posterior portion of Broca's area and the ventral premotor both control movement of the forelimbs, the face, and the mouth.

Consciousness does not require syntactical language. I posit that, since most people think and plan verbally, the evolution of syntactical language strongly favored higher states of consciousness that include self-reflection, thought, and the planning of actions. Concepts typical of human minds, including planning and such abstractions as "society" and "freedom," "exist" linguistically because we can talk about them. Various states of consciousness are of great selective advantage for individuals embedded in social networks (see chapters 3, 5, 7, 8, 20, 21, and 23). The idea of "Machiavellian intelligence" (Byrne 1995) hypothesizes that the social environment exerted a strong selection pressure that resulted in self-recognition, empathy, imitation, make-believe, deception, theory of mind, and teaching. These were used for social manipulation. The capacities to interact with the natural and the social environment, to make decisions and plans based on previous experience, and to modify behavior quickly in accord with new demands were all augmented with the rise of consciousness.

20
Brains and Symbols

John Skoyles

The word "chimera" refers not only to mythical beasts that combine various animal parts but also to animals that merge in one body populations of genetically distinct cells. Here Skoyles uses "chimera" in yet another sense—to refer to the blending of symbols with the human brain's neural apparatus.

A chimp cannot read this book, but you can. Nor can that chimp divide 7 into 20, promise to meet a friend, tell a joke, or deceive with words (lie). We humans are unlike any other mammals in certain ways. Why? It is not because chimps lack rudimentary math, a sense of friendship, humor, or an ability to deceive. Research finds that they are capable of these skills. But somehow we humans have expanded these beginnings of communication into entirely different abilities and affects—and then invent and keep reinventing them with fresh innovations. Reading, writing, and arithmetic all appeared recently in human evolution. From logographs, phonetic spelling, irregular spelling, to ancient religious inventories to blogs, language continues to evolve. Innovation relates to symbols and to the human brain's capacity to chimerize them, to recombine and modify them to produce something more than either brain or symbol. Further, not only do symbols change the nature of cognition and emotional affects; they also allow innovative speech and mental images to spread to other people.

This chapter is about how symbols manage this brain-change trick. The key to symbol chimerization is a process called "neural plasticity." This property of neurons had been dormant in brains for at least 200 million years, or perhaps even longer. But it took enlarged brains with prolonged development, in bodies able to easily communicate with others, before brains could exploit it to chimerize with symbols (Skoyles 2008a).

Neural Plasticity

We are a combination of many components, but a critical one is in the convoluted crumbled sheet of cortical neurons behind your eyes. This sheet is roughly 2,500 square centimeters in area, but only about 3 millimeters thick. On this sheet are written about 200 areas of cognitive skill per hemisphere. Different areas link to different tasks, a fact reflected in many of their names: Brodmann 17, primary visual cortex; Brodmann 41, primary auditory cortex; Brodmann 4, primary motor cortex; and so on. A major question in neuroscience is "How responsible are genes for various brain functions, and to what extent are various brain functions due to the environment?"

There are two alternative approaches: the "gadget" approach and the "blank slate" approach. According to the gadget approach, area-function links are written into cortical-area neural networks by genes picked and fixed for that purpose by natural selection. According to this mode of thought, the visual cortex is a neural gadget that receives input from the retina via the lateral geniculate nucleus to which neural "sight" circuitry responds. The ability to see is thus pre-formed by evolution into neural circuits. Just as we would not expect tissues that compose the eye to randomly join and form a complex optical device, according to this view, neither should we expect the brain to process vision without evolutionary preparation. The environment is seen but not directly involved in vision.

There is indeed evidence that the visual cortex co-evolved for light sensitivity and detection. (See chapter 11.) Neurons found nowhere else in the cerebral cortex are found there, with uniquely specialized circuitry. Moreover, neurons in the visual cortex, like those in other cerebral cortex areas, arise by cell division from particular areas in a stem-cell "protomap." They are retained even if transplanted elsewhere in the brain. Clearly, the various regions of the cerebral cortex were strongly shaped by evolution (Skoyles and Sagan 2002).

But "evolved does not mean fixed. What is written might be only in faint ink and thus easily rewritten. An alternative Darwinian "blank slate" approach argues that the brain is open to rewriting, as its neural circuits are capable of rewiring. Surely evolution shaped the brain and capacities exist in different areas for particular functions. Environmentally induced rewiring of brain circuits, such as the social and educational functions taught via symbols, became possible.

Are brain processes fixed, or are they flexible? The test is whether cortex that is not genetically written to process retinal input can visually recognize objects. It can. Auditory cortex that has not been selected to "see" can "learn" to process visual symbols if, early in development, this portion of the brain receives light, i.e., visual stimuli from the eye. Experimental animals in which half of the brain sees with normal visual cortex and the other half with auditory cortex wired to the eye act, when tested, as if there were no difference. The opposite mix-up happens in people born blind: with no eye input, the visual cortex can develop, and its neuronal connections can become auditory cortex. More remarkably, what would have developed as visual cortex in the brain can learn to process meaning and syntax of spoken words. Nonetheless, the "writing of genes" is also retained. Auditory cortex tricked into processing sight displays features that evolved for the mechanosensitivity associated with sound waves "Retrained" auditory cortex probably functions more poorly in light recognition than proper visual cortex. But that is not the point. We now know that evolution left the auditory cortex and other areas open, so that it is brilliant, in this case, at hearing and interpreting sound but at the same time does not forgo the ability (with the right environmental intervention) to be "retrained" as visual cortex.

Why did evolution not fix the brain? The heart, after all, does nothing more than pump fluid. Yet evolution left the auditory cortex capable of also processing sight. The reason for this flexibility is that it is a developmental by-product. Your fetal brain had three or perhaps four times as many neurons as your adult brain has. But neurodevelopment ("neural Darwinism"—see chapter 19) whittled them down by 70 percent. This "neural natural selection" also occurs in the development of axons and their synapses. The developmental credo in the nervous system is "Overbuild and eliminate excess." Such development generates brain functions by exuberantly pruning connected circuits into sparse but effective ones. "Effectivity" involves strengthening of synaptic densities. The number of these neuron-connecting thickenings that can be observed with an electron microscope may be 10,000 per neuron. As a consequence of this, during development the brain can be pruned down and can rebuild in radically different ways that produce new functions, including not only sight within auditory circuits but also such never-before-seen functions as the capacity to process complex symbols.

Neural plasticity is only one piece of the biology needed for the brain's processing of symbols. Spare cerebral cortex is also essential. Mark Twain once remarked that his capacity for promises had been crowded out by "other faculties" (Skoyles 2008b). And neuroscientists find he was right. After brain injuries, during recovery, faculties can crowd one another out. Thus, the symbol processing, if it is to exist, requires a large brain so newly developed functions will not crowd out old ones. We humans, as is well known, evolved very large brains with more cortical neurons and more neuronal connectivity than other primates have. (See chapter 19.)

The brain's use of symbols was accompanied by body skills—smart hands and smart vocalization. Our dexterous free hands and our specialized vocal tract and breathing respectively enable sign language as well as vocal language. Chimps are able to gesture, but that ability is of little use if their hands are used primarily for climbing or knuckle-walking. Moreover, although chimps make diverse sounds, they cannot string them together smoothly into complex vocal units to sing or talk. That requires a specialized ability to vocalize dozens of articulations continuously upon a single outbreath. In terms of respiration, it is a trick that even birds cannot manage: the notes that birds sing are separated by mini-breaths.

Language depends on ancestors. If you were born with normal vocal organs but were made to take part in an inhumane experiment in which all communication was barred, you would not, however, be able to use your voice, or even learn to speak. Already-existing input is essential. Language doesn't emerge *de novo* in each person; it is passed, with little modification, from generation to generation. We speak the language of our parents, they spoke the language of their parents, and so on, further and further back into the past.

Chimerization of the Human Brain

Language is not only communication. Non-communication brain function is also connected to language in many unexpected ways. Inner speech provides our brains with "sketch pads" on which to work out and organize ideas. By exploitation of written symbols, inner speech rewires even how we hear. Illiterates, more than literates, find it difficult to hear errors in speech or to identify the phonemes of spoken words. Rhyming is also affected by spelling: words appear nearly the same are more easily rhymed when read aloud by literate people. (Compare "there" and "where" with "tear," "ear," and "deer.") How we sense space or understand the sequence of events is partly determined by how our language expresses spatial and temporal differences. If early communication lacks reference to mind states, desires, intentions, and deceit (as is the unfortunate situation of deaf children raised by hearing parents that resort to lots of pointing), then awareness that others have inner minds is limited. Brain imaging reveals that spoken words affect even low-level unacknowledged perceptual decision processes (e.g., seek or avoid light) in the visual cortex. Our brains are chimerized with communication symbols—words, both written and spoken, that affect perception and behavior far beyond their use in communication.

Words, which in the past could only be heard by people nearby, now are written and stored. But writing forces the meanings of words to be more abstract and decontextual. Communication that is easily comprehended orally when words are aided by gestures and shared surroundings is not necessarily understood when it consists only of words. Writing, therefore, compels us to communicate without the aid of immediate and shared context. That written communication skill must make explicit what is otherwise left implied feeds back to improve the accuracy of how people speak, think, and behave beyond writing. However, written communication may also foster deception and profound misunderstanding.

Once an abstraction is linked to a written symbol, it can propagate, mutate, and accumulate complexity. In the words of Richard Dawkins (1990), "a meme evolves."

Brain-Symbol Chimeras in Evolution

The human mind arises from the brain and from symbolic culture. This happens through a chimerical fusion in which cortical circuits in the brain correlate with transmitted symbols. Chimerical brain symbols evolved as a result of the human expansion of the primate brain that provides extra circuit space for symbols and as a result of body innovations that enable the transfer of symbols. These innovations include bipedality (which frees the arms and hands for gesturing) and changes in the vocal tract and in respiratory air flow that make it possible to string together a rich variety of vocalizations. Such gestural and vocal corporeal memes enable symbols to hop horizontally and vertically between people with brains. Symbols rewire cortical circuits within the brain by exploiting a property called "neural plasticity." Think of "neural plasticity" as another name for learning. Symbols exploit brain faculties as they take up residence by co-opting ancient innate functions and rewiring them with new symbolic outputs and inputs. Residence of symbols in brains and exchange of symbols between brains underlie articulate thought and language and symbol-effected emotions and bonds. Brain and symbols have separate evolutionary trajectories. That of the enlarging human brain virtually ceased 180,000 years ago, but the evolutionary trajectory of symbols—words (written or spoken), flags, bar codes, and so on—continues to expand vigorously.

In a way, chimerization of the brain with symbols was as momentous for the rise of the mind, and for human-style consciousness, as enzymatic catalysts were for life. The oxygen, carbon, and other atoms of life that were spewed out by supernovas and ended up on planets do not replicate, respire, and evolve. They mostly just lie there across the universe as semi-organic gunk. Enzymatic catalysts changed that. They let this gunk evolve into cells that reproduce. Chemicals enzymatically organized themselves to generate large organic molecules that built organelles and started that automatic organization we recognize. (See chapter 2.) Chimerization let cells compile their talents to become even more complex. What were brains before symbols? They were raw perceptions, motor control, and emotions that were just sufficiently developed to enable continued animal survival. But now chimerized, they generate a rich diversity of attributes that had not existed previously. For example, mathematics was always possible in the universe, but now thanks to brains chimerized with mathematical notation it can thrive as a reality as human embodied brains prove theorems, analyze equations, calculate and compute numbers, and so give such mathematical possibilities life. Language is another case: it creates the possibility of stories and thoughts. Chimerization thus creates minds aware of imagined worlds (and even of themselves) because language gives them the tools to reflect upon the existence of a self with a continuing "I."

Symbols and Humanity

Symbols extended cognition, accelerating our tendency to become more cultural and social. Like other social animals, we cooperate and also engage in deceit, betrayal, and power struggles. (See chapter 7.) Humans chimerized the emotions that drive cooperative and competitive behavior with symbols. Culture, with its highly structured social coexistence, has expanded dramatically in the Holocene. Sexual and other bonding relationships became publicly defined (at least until modern times) in marriage. Some form of marriage, whether monogamous, polygamous, or polyandrous, is found in every human society. Most formalized relationships are rich in symbolic ceremony, ritual, and symbolic paraphernalia, some involving changes of names and forms of address, exchanges of jewelry, or scarification. Genetic (familiar) relationships became public knowledge and detailed with great refinement in extended kinship systems. Behavioral taboos became experienced as wrong and right, sin and virtue. Friendship, which in the past was expressed in primates mostly by shared foraging warnings and mutual grooming, found its existence reinforced by anniversaries, parties, and gossip. Other social animals belong to herds, troops, bands, castes, or clans, but only humans name these groups with vocalizations that extend over lifetimes. Such named social entities are bound by culturally shared emotions and symbols. We humans do not merely belong. We turn the words we speak, our gestures, and our clothing and the way we remove it into symbols of that belonging. We may be violent, but close study indicates that we are also extremely cooperative, with incessantly shifting alliances. (See chapter 7.) Because of the symbols that define us as group members, we work together or apart, even if the purpose of our work is to destroy people who use symbol systems different our own. As Fleck (1935) recognized, we all belong to "thought collectives" with distinctive "thought-styles." The thought collectives of scientists are instantly recognizable to scientists who belong and usually incomprehensible to those who do not.

Brain-Symbol Chimeras and Anthropology

The chimera–brain symbol approach suggests that anthropologists and neuroscientists should enter into dialog. The human brain needs to be studied more in the context of its extraordinary tasks, and not just by physiological methods.

Human social life is full of "don'ts," taboos, and rituals. One conjecture is that symbol chimerization, as prohibition admonishment, resides in the orbital prefrontal cortex and in other specific portions of the brain. Symbols chimerize their neural circuits by providing them with novel inputs and outputs. In apes, they underlie behavioral "don'ts." Symbolized, they may have made codified morals and etiquette possible. Human experience is full of magic, pollutants, unlucky numbers, evil eyes, sacred places, ritual ceremonies, superstitions. Fears emerge from the amygdala. Might processes in the amygdala have become signified through their symbols? Some people wear rings, powder, paint, or other tribal accoutrements. Do the insular or somatosensory cortex processes that link emotions to our bodies reside in specific regions of the brain? That is, does the emotional grounding of our symbols arise from exploiting already-existing feelings by linking them to artifacts, images, and other culturally determined experiences?

My brain-symbol approach, here called "chimeric," suggests that behind every anthropological account of cultural symbols there is a corresponding brain-chimera story. Culture is not separate from human physiology, including the brain and the peripheral nervous system. Rather, culture is an integral part of the nervous system's capacity for inventing and recognizing symbols.

The "brain symbol chimerization" perspective, therefore, challenges the present balkanization of neuroscience, human evolution, and cultural anthropology. Genes were selected not to build complex cognition but to enable brains and bodies to survive and leave healthy offspring. Bodies with brains gained competence from culturally transmitted symbols, made possible by brain expansion. Thus, although our brains are made of the same neurological components as those of other animals, they have experienced a radical innovation in that their cognition has come to depend on transmitted cultural symbols. The chimerization perspective suggests that human evolution, neuroscience, and cultural anthropology are a single phenomenon, to be studied together.

The chimerization approach also challenges scientific thinking that ignores how profoundly the brain is changed in its functioning by the culture in which it is embedded. The human brain is not merely an information-processing organ. Rather, the information processing of the brain depends on cultural symbols and on transmitted learning. Nor are we merely a symbolic culture; without living brains, symbolic culture, if it exists at all, is at most a collection of museum artifacts. We are what happens when human brains and symbolic systems—gestural, vocal and written—interact and enable each other's existence. Symbolic culture requires human brains and human function in the context of symbolic culture. However, human brains have not changed since the origin of modern *Homo sapiens*, whereas symbolic culture thrives 200,000 years after it began on a remorselessly accelerating and expanding trajectory of human transformation.

21
Thermodynamics and Thought

Dorion Sagan

Do consciousness and matter belong to radically different realms? Here Sagan argues that the sensitivities and responses of consciousness in humans and others emerge from naturally complex energy-driven systems.

Consciousness is a subjective datum whose existence in other subjects cannot be experienced directly. The inability to perceive the consciousness of others directly is a skeleton in the closet of philosophy, and is the phenomenological basis of solipsism and various forms of paranoia. It is also a source of profound reflection in the works of the science fiction writer Philip K. Dick. In these works, empathy, which allows the individual to connect in consciousness with another, is the defining mark of being human—a state which is, however, not confined to humans but may also be a property of "humanoid robots," as in the novel *Do Androids Dream of Electric Sheep?* and in the short story *Imposter* (Dick 1968, 1953). So, too, for Dick, people deprived of the ability to transcend their own phenomenological isolation and make an act of faith that others, too, are real, are inhuman. Perhaps these problems of the isolation of consciousness conscious of its own isolation perhaps do not trouble other animals, which are connected via instinct and emotion with less capacity for imagination conferred by the intellect. In general, we attribute internal perception to organisms, and only organisms, whose physiques and behavior resemble our own. Thus, another person will be deemed conscious, a dog less so, and an amoeba perhaps not at all. A basic assumption of metaphysical materialism, also known as naive realism, is that consciousness emerges from the interaction of parts. This cannot, however, be proved, since consciousness—whether defined as awareness, as awareness of awareness, or in some similar way—comes from the inside. Ironically, some theorizers of artificial life have used the intrinsic inability to detect awareness as an argument for the possibility of its existence in computer programs. The famous Turing Test involves a person guessing whether an entity is alive from its answers alone, without knowing ("seeing") whether the source is a mechanical computer or flesh and blood. In theory this essay could have been written by a machine, though I doubt that any machine is capable of such nuance and simulation. The great scholar and historian of Gnosticism Hans Jonas argued persuasively in his 1979 book *The Phenomenon of Life* that scientific materialism—a world without spirit in which minds are the result of brains and brain-like connections of particles—cannot be understood without looking at the history of science. In this history, the philosophy of Descartes is crucial. Whereas, Jonas says, our ancestors believed that anything that moved was imbued with a feeling spirit, experience taught them that was probably not so. The origins of Western science in Greece dovetailed with skepticism about the status of spirits and gods.

Descartes advanced this ostracizing of spirit from the natural world for the sake of science by identifying thought and feeling with the pineal gland, which was supposed to be connected to God, a sort of hotline of spirit between humans and their maker in an otherwise mechanistic world. This untenable move, although belied by the subsequent discovery of pineal glands in other organisms, expedited science after Newton, giving it the equivalent of a permit or "license" to consider anatomic and earthly bodies, no less than astronomic and celestial ones, mechanical creations. As Jonas underlines, however, at no point during the Cartesian dismantling of living beings do we ever come across any part, or any combination of parts, that might, even in principle, confirm the existence of consciousness. That is because the spirit-matter dualism from which science's materialistic monism derived was not superstition but the result of empirical observation of inner states by subjects unable, except by empathy and the circumstantial evidence of likeness, to verify the existence of such states in their kin. This is the "ghost in the machine" without which we ourselves, explained as merely mechanical creations, are explained away. No theory of consciousness is needed to explain the behavior of zombies, or that of unfeeling humanoid robots. But we are not zombies or robots—at least, I am not. Thus, as Jonas argues, dualism remains alive and well beneath the surface of a thoroughgoing naive realism or metaphysical (scientific) materialism that assumes mind's emergence from matter and energy.

The inability of perceiving beings to establish connections outside themselves is the meaning of the famous statement that "the monads have no windows" (Leibniz 1714) and the efforts of Varela and Maturana to highlight organisms' informational closure as self-referential beings continuously producing themselves. Gregory Bateson's 2002 book *Mind and Nature* also points up the belonging-together of consciousness and physical systems, even as it continues to sharply distinguish between them. David Bohm (2002) argues that minds and bodies, far from being reducible to each other, are both projections of a deeper underlying reality. Although not quite opposites, mind and body, or spirit and matter, represent a classical binary hierarchy of linked metaphysical terms one term of which is necessarily privileged (currently, in science, matter) at the expense of the other (currently, mind). The philosophical-rhetorical method known as "deconstruction" entails the "reversal" of such hierarchies, followed by their "displacement" (Derrida 1982). Culturally, we have vacillated between idealism's "everything is in the mind" (perhaps of God) and metaphysical materialism's (naive realism's) "it's all matter." This is the "reversal" of the mind-matter split. For "full" deconstruction it would not be necessary to "displace" the persistent metaphysical opposition.

Freedom without Free Will

Spinoza, in *The Ethics*, ascribes the scientific realm of determinist interactions—the world of scientific mechanism—to human consciousness. Spinoza is a kind of "Cartesian monist": he extends the mechanical realm of the *res extensa*, which Descartes lavishly extended to most of creation, to human mental processes and God; at the same time, he extends to all of creation the *res cogitans* that Descartes had isolated to the realm of human thinking and a transcendental God. Mind and matter still existed, but now they overlapped. The causal, mechanical behavior of caprice-free scientifically studiable objects now applied to human mental processes, despite the appearance of free will. Nor could God be excluded from the infinite chains of cause and effect enveloping nature as revealed by science. In retrospect, we may say that Descartes completed the displacing deconstruction of the mind-matter split whose hierarchical relationship (mind over matter) Descartes had, in a preliminary movement, reversed.

Thus, Spinoza goes Descartes one better, "pushing the envelope" of the extremely expeditious move by which Descartes made the world, especially the world of organisms, safe for scientific investigation. But in following and ratcheting up Cartesian materialism, mechanism, and determinism—by philosophically making science, as it were, more scientific by eliminating the realm of divine caprice from God—Spinoza dealt the notion of free will a death blow. In a letter explaining his views to two correspondents, Spinoza writes:

I say that a thing is free if it exists and acts from the necessity of its own nature alone, and compelled if it is determined by something else to exist and produce effects in a certain and determinate way. For example, even though God exists necessarily, still he exists freely, because he exists from the necessity of his own nature alone. So God also understands himself, and absolutely all things, freely, because it follows from the necessity of his nature alone that he understands all things. You see then that I place freedom not in free decree, but in free necessity.

But let us descend to created things, which are all determined by external causes to exist and to produce an effect in a certain and determinate way. To clearly understand this, let us conceive something very simple—say, a stone which receives a certain quantity of motion from an external cause which sets it in motion. Afterward the stone will necessarily continue to move, even though the thrust of the external cause ceases, because it has this quantity of motion. Therefore, this permanence of the stone in motion is compelled, not because it is necessary, but because it must be defined by the thrust of the external cause. What is to be understood here concerning the stone should be understood concerning any singular thing whatever, no matter how composite it is, and capable of doing a great many things: that each thing is necessarily determined by some external cause to exist and produce effects in a certain and determinate way.

Next, conceive now, if you will, that while the stone continues to move, it thinks, and knows that as far as it can, it strives to continue to move. Of course since the stone is conscious only of its striving, and not at all indifferent, it will believe itself to be free, and to persevere in motion for no other cause than because it wills to. And this is that famous human freedom which everyone brags of having, and which consists only in this: that men are conscious of their appetite and ignorant of the causes by which they are determined.

So the infant believes that he freely wants the milk; the angry boy that he wants vengeance; and the timid, flight. Again, the drunk believes it is from a free decision of the mind that he says those things which afterward, when sober, he wishes he had not said. Similarly, the madman, the chatterbox, and a great many people of this kind believe that they act from a free decision of the mind, and not that they are carried away by impulse. Because this prejudice is innate in all men, they are not easily freed from it. (Curley 1994)

Einstein followed Spinoza in believing, partially on deistic grounds perhaps, that everything is caused, but not by will. In a famous letter to Max Born, Einstein declared that God "does not throw dice."

Spinoza, despite his denial of human free will, was an intense advocate of the freedom to worship and of speech, and his ruminations are part of our cultural backdrop. Spinoza appears to have believed, however, that freedom was ultimately equivalent not to free will, but to knowledge. And he believed that, because God, of which matter and mind (Descartes' *res extensa* and *res cogitans*, respectively) were infinite parts accessible to us within an infinite being with infinitely more parts, was eternal, God was not free. Freedom entails decisions made in time, but a Being outside of time need not and cannot exercise choice or free will, since all would-be "decisions" have already always been made. However unidirectional events may appear, for Spinoza this "prejudice is innate" and in reality we dwell in a completely closed causal network for all time.

Purposeful Behavior in Inanimate Systems

Thermodynamics is the study of energy and its transformation. Initially, thermodynamics focused on systems closed to both matter and energy exchange ("isolated systems"). Such sealed systems inevitably reach equilibrium, a steady state of relative randomness and exhaustion from which no work can be derived. The idea that intelligence might be able to extract work from such "final" states was the point of Maxwell's famous "demon" (Leff and Rex 1990). The most important observation of classical thermodynamics is that heat flows from a hotter to a cooler body—in other words, that energy spreads. Although the second law was stated in terms of entropy (originally heat divided by temperature, but later given a statistical formulation), its modern statement, which applies to complex systems as well as to simple systems, is that energy, if not impeded, spreads or, in the words of the ecologist Eric Schneider, that "nature abhors a gradient" (Schneider and Kay 1989). A gradient is simply a difference across a distance, such as a difference in temperature, pressure, or chemical concentration. Gradients mark regions of energy flow. The thermal difference between a hot boiler and a cold radiator drives a steam engine, for example. This is a thermal or temperature gradient. A tornado forms between high-pressure and low-pressure air masses. This is a barometric air pressure gradient. Perfume spreads from a woman's neck to the rest of the guests at a table; this is a chemical gradient. In both the case of the human-made machine and the natural complex system, complex structured processes run on the flattening of gradients or the spread of energy. In fact, the appearance of multiple typhoons and life's biospheric expansion both reflect (rather than contradict) energy's second-law-described mandate to spread. It is, thus, not true that the second law mandates disorder: the complex, naturally organized tornado and the highly nonrandom steam engine are both driven by the energy efficiencies of the second law. The modern versions of the second law apply to open systems as well as to thermally sealed systems. Moreover, they are easier to understand, and they dispense with the need to include entropy, an elusive quantity that is difficult to picture because it is a ratio.

Modern thermodynamics chronicles how organized structure can be maintained and can grow in areas of energy flux. In fact, complex, organized cycling systems appear to be more effective at producing atomic and molecular chaos than random distributions of matter. Consider a "tornado in a bottle," a whirlpool formed when a 1.5-liter soda bottle filled with water is inverted and attached to another such bottle with a small column of punctured plastic. The system drains more quickly via organized cycling than via slow dripping. Such cycling may be considered an elegant rectification of a potential energy gradient. The complex system, the cycling whirlpool of air or water, or the chemically based, metabolically cycling system of life effectively dissipates preexisting gradients. The complex organization spreads energy and produces entropy faster, or more stably, in the end spreading more energy than would have been spread if it didn't exist. Heat, entropy, and reacted-out waste gases are exported to the far domains of the energy-transforming systems whose tendency to grow reflects the second law restated as energy's tendency to spread. Life and other natural complex systems, far from violating the second law, appear actively to accomplish the "goal" implicit in equilibrium as an end state. The purpose of life and other naturally organized complex systems is integrally connected to time's arrow as defined by the second law and energy's spread. The swirling forms that spontaneously appear when conditions allow have a natural function. They are the efficient cause of reducing gradients.

Purposeful Life

As with Spinoza's conscious stone, the end-directed behavior of non-living systems to reduce ambient gradients seems to have evolved in living beings as purposeful behavior and in humans as the illusion of conscious purpose. Conscious beings, modeling themselves and their surrounding, are equipped to seek out the energy gradients on which they depend. Prokaryotes with rotary flagella, both archaebacteria and eubacteria, swim up chemical concentration and light gradients, literally toward sweetness and light. Tornadoes, hurricanes, whirlpools, dust devils, lithospheric plate motions, the northward flow of the Gulf Stream, Bénard (convection) cells, Taylor vortices, Liesegang rings, marine salt fingers, atmospheric Hadley cells, photosynthetic bacteria, slime molds and other colonial cell forms, the continuous "whirlpool" downstream of Niagara Falls, gradient-feeding metabolizing proto-life, local economies, rainforests, you, the planetary ecosystem, panthers, Jupiter's Great Red Spot, Red Spot Jr., and Baby Red Spot, and perhaps even spiral galaxies and the helical shape of DNA all reflect the dynamics of energy-driven cycling systems. Chemical "clocks," such as the Belousov-Zhabotinskii reactions, reduce chemical gradients as they generate waves across complex chemical structures. Although such systems do not replicate, they do maintain and grow more complex. Because nature "selects" at a pre-biological level for complex systems that spread energy, that is, for "stable vehicles of degradation" (Wicken 1987), it seems clear that replication would have aided, and eventually helped stabilize, complex systems whose normal function, to reduce ambient gradients, could now be prolonged (Sagan and Whiteside 2004). As such systems maintained and developed, their abilities to construct themselves continuously (always connected to the destruction of the gradients around them) and to recognize the gradients on which they depended for their livelihood and growth would have tended to improve. The conscious awareness Roth describes in chapter 19 abets our access to gradients and helps perpetuate our living maintenance of gradient-reducing systems. Our fear of death, for example, correlates with a lust for life that tends to increase (or at least accompany) our persistence as gradient-reducing systems.

Foraminifera select and discard different-sized and different-colored glass beads for use in constructing the agglutinated "tests" that form their colorful external skeletons. Such proto-conscious "choice" in the agglutinaceous marine protoctists represent the sort of basic physiological infrastructure from which consciousness "proper" (that is, as we know it in ourselves) probably evolved. Physiological organs (e.g., the sex organs), sometimes said to have a "mind of their own," anticipate consciousness in that they clearly display functions that are not usually contested in biological discourse because they have obvious mechanical explanations. The function of the heart is to pump blood, that of the kidneys to purify the blood, and that of the eyes is to see. Such functionality is explained as the result of natural selection among living things. But even non-living systems that do not reduce gradients tend to disappear, as they have no natural, sustainable function as agents of degradation.

The basic purpose of living beings—the purpose from which the variegated physiological organs arise during the course of bacterial, protoctist, and later animal, fungal, and plant evolution—is to degrade gradients. Thermal and satellite measurements show that the more complex ecosystems (e.g., rainforests versus deserts, temperate woodlands versus arctic seas) are more effective at producing heat (entropic waste) as they keep themselves cool. James Lovelock has shown that the simple behavior of light-reflecting and light-absorbing black and white daisies growing within the same specified temperature range (4–40°C) can, in principle, produce planetary thermoregulation—an "optimizing" process. Often dismissed (despite evidence for it) in the case of the whole global ecosystem, environmental temperature regulation is denied because it seems to indicate physiology or, worse, conscious purpose at the planetary scale. In fact, our gradient-oriented behaviors and feelings of purposeful will probably should be understood not as an illicit ascription of conscious, human-like purpose to planetary processes, but rather as closely related to the end-directed behaviors described by the second law of thermodynamics.

Complexity in nature, which we easily and naturally associate with the product of our own minds, requires neither human-like creation epitomized by the mechanical process of putting things together from constituent parts nor natural selection among living systems. Rather, seemingly designed structures and processes spontaneously appear to reduce gradients and spread energy in accord with thermodynamics' second law. Purposeful, mind-like, highly directed, and circuitously organized behaviors thus need not be divine, if by divine we mean conscious in a human way. Much of what passes for anthropomorphism reflects the Cartesian privileging of matter over mind: because the *res extensa* was not extended far enough into human cognitive processes, let alone the *res cogitans*, which was severely delimited in its extent, the effect of recognizing mind-like behaviors in nature is to elicit accusations of pathetic fallacy, personification, and superstitious pre-scientific thinking. The truth is the exact opposite. Once the Spinozistic program is accepted, and the mind-matter metaphysical hierarchy is deconstructed beyond the preliminary "reversal" stage, the taboo against recognizing teleological processes in nature disappears. This is salutary. The alternative to an unscientific anthropocentrism that rigidly separates humanity's thoughtful, purpose-driven, and aware activities from the rest of nature is an evolutionary connectionism that sees us as an integral part of life and the rest of nature. This is a "reverse anthropomorphism" in which the human-like attributes we detect in nature are not bogus exports from us to nature, but rather reflect that purposeful, mind-like complexity exists in end-directed, second-law-driven nature in the first place.

Ironically, Aristotle—whose insistence that organisms did not develop randomly but rather toward a telos, or end, was embraced as proof of a human-like God by the Christian Church—wrote the following more than 2,000 years ago:

> In natural products the sequence is invariable, if there is no impediment. . . . It is absurd to suppose that purpose is not present because we do not observe the agent deliberating. Art does not deliberate. If the ship-building art were in the wood, it would produce the same results by nature. If, therefore, purpose is present in art, it is present also in nature. . . . It is plain then that nature is a cause, a cause that operates for a purpose. (McKeon 2001)

Aristotle defends teleology as inherent in living beings. (See II.1 and II.8 in McKeon 2001.) He likens the purposefulness of organisms to a doctor who happens to treat himself: although made of matter (everybody can be a patient), organisms work on themselves (like a doctor operating on himself). Because for Aristotle the organisms of nature together seem to be the only example of the phenomenon they display, it is difficult to find analogies for their peculiar behavior. But Aristotle says his self-treating doctor is "the best example." Because art often imitates nature, and sometimes completes what nature starts, Aristotle is also interested in looking at art so as to understand organs that seem designed for a purpose but are inherent in the organisms growing them. In animals, Aristotle argues, front teeth and molars do not occur simply by chance; they serve two discrete purposes: cutting food and grinding it before digestion. Similarly, plant roots do not just happen to move down into the ground; they orient themselves that way for the purpose of nourishment. Aristotle argues that the parts of non-living things are different than those of living things, since living things are purposefully organized. He even goes so far as to say that, so far from being reducible, living beings are prior to their parts. This makes sense thermodynamically when we consider that the *ur*-purpose—the original goal—of the systems we call living is to garner energy for acts of survival that inevitably reduce chemical, light, or other gradients accessible to the living. The downward-seeking of roots, the grinding teeth of animals, and the bending of plants toward light are long-evolved offshoots of an evolutionarily ancient function. That function, or purpose, is to reduce gradients in accord with the second law and simultaneously live to reproduce. A free God intervening willy-nilly in nature is as incompatible with a scientific cosmos governed by eternal mathematical laws as it is with a picture of life's purposeful, energy-based behavior emerging from inanimate thermodynamic roots. We may speculate that consciousness, far from belonging to a radically distinct "spiritual" realm, is an outgrowth or variation of the naturally complex behavior of self-maintaining thermodynamic systems.

22

"I Know Who You Are; I Know Where You Live"

Judith Masters

If we can reasonably speculate that consciousness as we experience it in ourselves is at least partially, if not entirely, the emergent result of the aggregate feelings and interactions of post-bacterial cells, it behooves us to turn our attention to the "mental capacities" of groups of animals—behaving, solving problems, and setting and accomplishing goals as collectives. Here Masters follows jungle primates that keep track of one another in dispersed social networks.

Living primates are classified into two suborders: the Haplorhini (literally, simple nostrils), which includes tarsiers, monkeys, and apes, and the Strepsirhini (twisted nostrils), which includes the lemurs of Madagascar (see chapter 15) and the lorises and galagos of Africa and Asia. All haplorhine primates, with the exception of Southeast Asian tarsiers and South American owl monkeys, are diurnal and gregarious. They have well-developed visual systems with binocular (stereoscopic) vision and notable visual acuity. The foveas in their retinas have a high percentage of cones (modified cilia, see chapter 13), which allow color discrimination. All African and Asian monkeys and some New World monkeys have trichromatic vision: they have three kinds of photopigments with peak wavelength sensitivities in the blue, green, and red regions of the color spectrum (see chapter 11). Thus, their communication repertoires often include striking visual signals. For example, the callitrichines of South America and the cercopithecoids of Africa and Asia use strongly contrasted color patterns to signal their species identity. Our own diurnal habits and visual sensitivities generate instant color skin recognition and pelage (fur) differences. They are signals of taxonomical significance.

Some of the lemurs on the island of Madagascar (e.g., *Varecia variegata, Hapalemur aureus, Eulemur coronatus*, and members of the

E. fulvus species group; see chapter 15) have incorporated color into their signaling systems, and this too correlates with diurnal activity. Although most of the larger lemurs have been described as diurnal, closer study has shown that many are active both by day and by night, an activity pattern we call *cathemeral* (Tattersall 2006; chapter 15 above). These lemurs may lack a typical tapetum or any tapetum at all (e.g., those in the genera *Varecia* and *Eulemur*). *Lemur catta* and *Hapalemur griseus* have rudimentary foveas with both cone and rod receptor cells (Pariente 1975). The cones are much sparser than in haplorhine primate retinas, and color discrimination in lemurs ranges from exceedingly poor to fair. The quality of the sensitivity and the response to color depend on light intensity. Trichromatic vision may exist in *Propithecus* and *Varecia* (both diurnal) and in the nocturnal *Cheirogaleus*.

The majority of lemurs, however, like their relatives on mainland Africa and Asia, are active only at night. In a dense forest, the canopy filters 99 percent of ambient light (Charles-Dominique 1977). The visual systems of nocturnal strepsirhines have evolved to detect this paucity of light by hypertrophy of the eyeball and the development of a reflective tapetum behind the retina that amplifies whatever incident light impinges on the retinal rod cells. The absence of a fovea and the virtual absence of cones do not preclude the use of vision in the perception of the nocturnal environment of these primates. They move around in relative darkness, some by rapid, almost ricochetal locomotion; they leap from one vertical support to another with great accuracy. However, the primary senses used in nocturnal foraging are olfaction (chemosensory) and hearing (mechanosensory) via kinocilia and stereocilia of the inner ear and balance organ (see chapter 13). They are crucial, too, for intra-specific communication.

Body odor is extremely important to the strepsirhine way of life, and strepsirhines are distinguished from the haplorhine primates by the development of two adaptive features used in the grooming and maintenance of the fur. Living strepsirhines (with the exception of *Daubentonia*, the aye-aye) share a modification of their lower anterior dentition: the toothcomb. The long, attenuated canines and incisors are rotated horizontally. The absence of functional canines has led to the evolution of caniniform anterior premolars. Under the tongue is a cartilaginous sublingua with fringed edges or lappets that are inserted between the tines of the comb to dislodge detritus that collects there. Additionally, the nail of the second toe is modified as a grooming claw, used to remove debris from the fur. By grooming themselves and their neighbors, strepsirhine primates spread their own scent throughout their own fur and transfer it to their closest associates.

The lack of emphasis on visual signaling systems in nocturnal strepsirhines means that species identity is generally not signaled morphologically as it often is in haplorhines. Slight variations in fur color will not be detectable under the low light intensities of a nocturnal forest. Species identity is signaled by odor or by vocalizations, characteristics that are invisible to taxonomists working on collections of skulls or stuffed skins in natural history museums (Masters 1993). My research into nocturnal strepsirhine diversity thus led me to focus on their communication signals. Here I discuss some of my findings regarding the olfactory and auditory signals emitted by greater galagos (*Otolemur*), the largest of the African bushbabies.

Galago Communication

The nocturnal strepsirhines tend to forage alone and are often thought of as solitary animals. Yet examination of their sleeping associations reveals complex social networks mediated by auditory and chemical signaling (Charles-Dominique 1977; Bearder 1987). Although there is little active defense of individual territories, each animal confines its activities to a "home range" that overlaps with the home ranges of its neighbors. The areas of overlap often include favored sleeping trees and particularly productive resources (e.g., fruit trees). The boundaries of the home range are marked by urine and by scents produced by specialized glands. The ranges of females are generally more restricted than those of resident males: an adult male's range may overlap with those of three or four females. Contact between males and females is maintained throughout the year, not limited to the mating season, and social bonds are reinforced by visits, by vocal communication, and by scent marking. A female is relaxed when she occupies the same area as the male to which she is bonded, but becomes wary and moves quickly through areas occupied by unknown males. Males behave similarly when they encounter evidence of the presence of unknown females.

During the mating season (two to three weeks in July, the austral winter), adult males become both more vocal and more mobile, ranging further to encounter more females. Mate preference appears to depend on familiarity, and estrous females spend most time with the resident males to whom they have become bonded. There is no paternal care of the offspring. Mothers and daughters tend to form stronger bonds than mothers and sons; whereas a son leaves the natal range upon reaching adulthood (at about 9 months), a daughter may share a large part of her mother's range for a lifetime. The average period of tenure of resident males is unknown; presumably they are replaced periodically by younger males.

The galago communication signals most readily accessible to humans are their loud vocalizations, the source of their common name: "bush-babies." Each galago species emits a powerful long-range vocal signal, capable of transmission over 3 kilometers (Petter and Charles-Dominique 1979). The call, most often emitted by males, acts as a "gathering" signal at the beginning or end of activity periods (at dawn and dusk). Similar, highly species-specific calls are given by lemurs, mangabeys, guenons, gibbons, and other primates (figure 22.1).

To assess their information content, I analyzed loud calls given by greater galagos (otolemurs) (Masters 1991). Both *Otolemur crassicaudatus* and *O. garnettii* emit calls that consist of a series of repeated motifs or sound pulses (figure 22.1). Such calls are particularly appropriate for long-distance communication, as the attention of the recipient is alerted

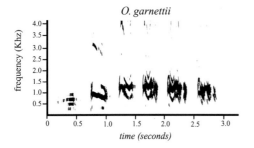

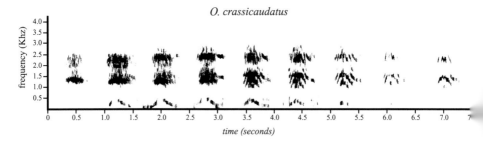

Figure 22.1
Loud calls of *Otolemur garnettii* (above) and *O. crassicaudatus* (below) (sonograms).

by the first motifs. The most obvious species-specific difference is in overall duration: calls of *O. crassicaudatus* are more than twice the length of those of *O. garnettii*. *O. crassicaudatus* calls generally contain a greater number of motifs, and each motif and inter-motif interval lasts longer than its counterpart in the call of *O. garnettii*. *O. garnettii* lives in the dense tropical dune forests of East Africa, and its calls are more modulated than those of *O. crassicaudatus*, which frequents more open riverine and coastal forests in southern Africa. The fundamental frequencies of the two calls are similar (between 1,000 and 1,500 hertz), which reflects similarities in their environments. Because frequencies above this level are scattered by forest leaves, it is more difficult to locate the source of the call (Waser and Brown 1986). The most consistent aspects of the call structures (and, therefore, those most likely to encode species identity) are the dominant frequency bands and the durations of individual motifs and inter-motif intervals. Individual animals can be recognized by their harmonics and timing, e.g., the frequency and overall length of their calls.

Besides vocal repertoires that provide instantaneous information to listeners regarding the identity and position of individuals in the community, galagos produce and respond to olfactory signals that yield similar information over longer time periods. Many strepsirhines dribble urine onto one palm and wipe it against the sole of the foot on the same side of the body, then repeat this with the other foot and palm. The scent of their urine is then deposited everywhere they move (Charles-Dominique 1977). Male bushbabies rub their cheeks in the urine marks of the females they are courting. Courtship bonds are established through long bouts of mutual grooming as the urine on one partner's hands is deposited on the other's fur. Social bonds established this way persist even after three months' separation in captivity. When the partners are reintroduced, they immediately recognize one another and begin extensive grooming. Other odors from specialized glands are also exchanged during this process.

Adult greater galagos of both genders bear on their throats and upper sternal regions a patch of glabrous skin on which droplets of bright yellow liquid are seen. The "sternal gland secretion" is transferred to substrates by "chest rubbing." A galago, most often an adult male, wraps its arms around a branch and pulls itself forward in a sliding motion. The anogenital region may contact the branch as part of the same motion. Sternal gland secretions serve a territorial function: "marking posts" within the home range are rubbed on repeated occasions, both by residents and by itinerant galagos. During the mating season, a male may approach a female and push his nose down between the female's arms to sniff or lick the sternal gland secretion.

Sternal gland secretions apparently contain information regarding gender, sexual maturity, and individual and species identity. Chemical analyses of the secretions produced by *Otolemur crassicaudatus* and *O. garnettii*, including three subspecies of *O. crassicaudatus* (*O. c. argentatus*, *O. c. monteiri*, *O. c. umbrosus*), consistently revealed the presence of three major organic components: benzyl cyanide (bc), 2-(p-hydroxyphenyl) ethanol (p-hpe), and p-hydroxybenzyl cyanide (p-hbc) (figure 22.2) (Crewe et al. 1979). The three components differ in volatility and hence provide information about the time that has elapsed since deposition of the scent. Bc evaporates within an hour or two; p-hbc and p-hpe evaporate more slowly. Waxes and oils in the sebum modulate the evaporation rates. Sternal gland scent marks need to be renewed every three to four days (figure 22.2).

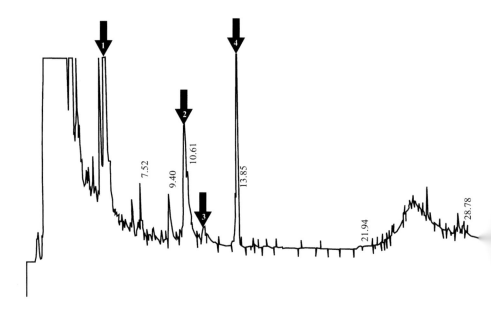

Figure 22.2
Gas chromatogram of the sternal gland secretion of an adult male *Otolemur crassicaudatus umbrosus*. The peaks represent the following chemical constituents: (1) benzyl cyanide, (2) p-(hydroxyphenyl) ethanol, (3) p-hydroxybenzyl cyanide, (4) n-pentadecane (internal standard). The concentration of peaks on the right includes compounds with high boiling points (oil and waxes), which provide the matrix for, and may regulate release of, the messenger chemicals.

The constancy of composition within species and between related species implies that the major message conveyed is not species identity. Rather, these chemical mixtures communicate more idiosyncratic signals that probably permit community members to track one another's movements along foraging routes. Association of a sternal mark with an anogenital mark allows a galago to advertise its sexual status along with its individual identity. "Marking posts" therefore enable information exchange, orientation of individual movements, reinforcement of pair bonding between males and females, and the integration of social and reproductive behavior in primates that are not overtly gregarious. This supports the contention of Beauchamp et al. (1976):

Individual recognition may turn out to be the major functional role played by mammalian secretions and excretions. That is, under natural conditions, who the individual is rather than what class it belongs to may be the question of greatest moment to the animal.

Conclusions

Specific mate recognition is essential to sexual reproduction in animals. In the case of galagos, recognition is based primarily on odors and vocalizations, although visual signals probably are significant at close quarters. My analysis of greater galago communication signals and behavior allows me to draw two major conclusions. First, the signals encode at least two levels of information: species and individual identity. Second, mate recognition takes place well in advance of mating. Generally, one associates animal courtship behavior with the act of copulation. Among primates, however, courtship serves to establish social bonds between individual males and females that may not mate until later—perhaps months or years later. Unattached females may be courted before they reach puberty, when lactating, when gestating, or in any other phase of the life history. Even in strepsirhine primates, many of which have relatively diffuse systems of social organization, courtship social behaviors extend beyond the brief act of copulation.

Acknowledgments

My attendance at the workshop in Bellagio would not have been possible without the support of Lynn Margulis and considerable assistance from Celeste Asikainen, and I express my heartfelt thanks to both of them. Thanks are also due to my fellow participants in the workshop for sharing their insights, and to my colleagues Max Del Pero, Laurie Godfrey, and Mike Mostovski, who played no small role in getting me there. Fabien Génin and Hugh Paterson provided generously of their time in improving an earlier version of this manuscript. My research is supported by the National Research Foundation under grant 2053616 (Conservation and Management of Ecosystems and Biodiversity). Any opinions, findings, conclusions, or recommendations expressed in this material are those of the author, and the NRF does not accept any liability in regard thereto.

23

Cultural Networks

Luis Rico

Working and living together on a crowded planet full of sensation and intelligence are inescapable facts of life. The way forward is a road not only of species divergence but of evolutionary chimerization: coming together to become more than any of us can ever be alone. Here Rico argues the importance of education of our true, chimeric nature for art, science, and other aspects of culture.

Art, science, and other modes of knowledge interact in the emergent paradigm of the "network society" that Manuel Castells (1996) described as a social system powered by microelectronics-based information and communication technologies. Current artistic, scientific, technological, and even commercial designs share tools and practices of mutual influence. Their strategies co-evolve in ways, largely undescribed, that extend beyond the language, religious, national, political, and disciplinary boundaries that arbitrarily, and often unconsciously, restrict knowledge.

Culture arises out of biology as interactive webs and networks. Systems of communication in the microbial world were already robust in the Archean eon. Networks, and even complex communities, began in the microbial world, and were amplified and extended among protoctists such as *Labyrinthula marina*, long before animals and plants evolved (Margulis and Sagan 2007; figure 23.1, left). Microbial nets have expanded, complexified and grown ever since. (See chapters 6–8.)

We humans have begun to view ourselves less as a landscape surface feature on the landscape that progresses linearly through time and more as a dynamic network of linked points on a tangle-prone web (figure 23.1, right). Educational institutions are divided into entities: discipline-limited departments, bureaucracies, and other groups with specialized, well-defined, generally temporary tasks. Empowered individuals (bosses,

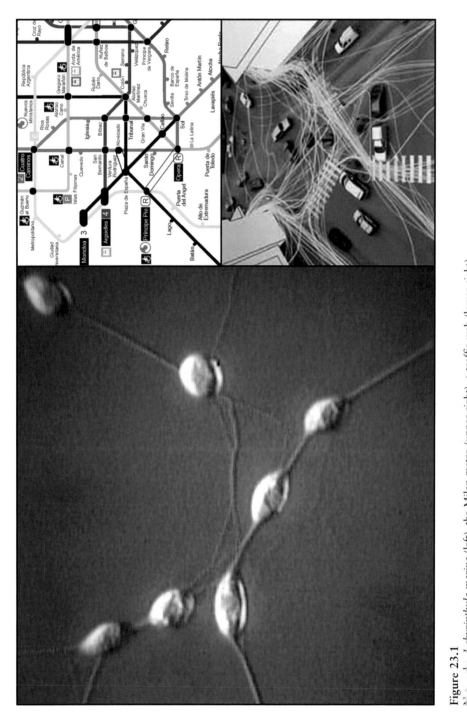

Figure 23.1
Physarum polycephalum moving (left), the Milan metro (upper right), a traffic web (lower right).

chairmen, directors) jealously defend their assignments that, for their own convenience, impose social relationships on their subordinates (workers, committee members, artists, researchers). Insensitivity to social issues or displays of rank often generate contentiousness and codified misunderstandings that interfere with innovation, see chapters 7 and 8.

Prodigiosity in science and in art emerged in interdisciplinary dialogs; interaction of separate and disparate traditions led to cultural novelty. Many great scientists and artists were holistic thinkers: Leonardo da Vinci, Charles Darwin, Albert Einstein, Galileo Galilei, Wolfgang von Goethe, James Hutton, Johannes Kepler, and Isaac Newton come to mind. Not only are "science" and "art" activities inseparable from the cultural (especially the national and economic) context that generates them; they are also inseparable from one another—as categories they are artificial. For example, the drawings of Santiago Ramón y Cajal, who discovered and gorgeously depicted neurons as units of the nervous system, represent creative expression that integrates discipline, sensibility, and scientific rigor (figure 23.2).

Fragmentation

New perspectives in many fields of knowledge sometimes are unwittingly let loose upon the world—one example is the spectacular NASA Apollo "blue marble" image of Earth from space. Such surprises speak to us directly; they transcend the cultural labels of art and science. Through this dazzling image we were startled into recognition of our limitations. Recently evolved African primates, we are wily, weak, small, vagabond, curious, chatty, but barely mutually intelligible. Our species colonized remote South Pacific islands and mountainous snowfields only during the Holocene (the last 10,000 years), at the very end of the Pleistocene epoch, which began 1,800 million years ago (Sagan 2007). As can be inferred from that photograph, all of us inhabit a small pale dot.

Our modes of learning influence how we perceive and relate to one another and to our habitats in the greater environment. Science, art, and technology generate awareness: the sub-visible and the emotional become apparent; pattern and process become explicit as boundaries blur. Life comprises interactions of complex networks, relationships that undergo continuous change at differing spatial and time scales. Such change in complexity favors new ideas, activities, musical instruments, measurement tools, and the like that mandate new human, material, and energy flow relationships.

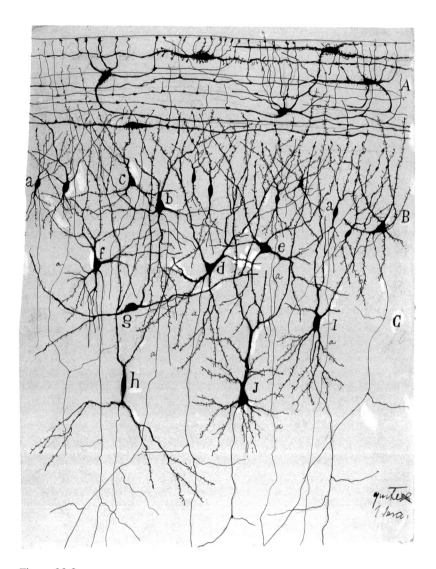

Figure 23.2
A drawing by Santiago Ramón y Cajal depicting neurons, units of the vertebrate nervous system.

Deterministic and dichotomous viewpoints labeled "science," "art," "religion," and "technology" insufficiently embrace real-world complexity. Details are routinely excluded by such inadequate canonical classifications. Professional academics especially dismiss nearly everything relevant as "out of one's field of expertise." Such gross communication failure fosters personal and international crises. The malaise, reflected by incessant exploitation of the living environment, affects everyone. Nets of communication (satellite, telephone, television, Internet) expose the consequences of local abuses and greed that underlies global change. We are "in the same boat." David Bohm, an expatriate of the United States who taught physics and philosophy in Latin America and in England for many years, wrote:

> The true crisis does not consist of the facts that confront us, but of the thought that has engendered them. . . . One of the difficulties of thought is its tendency to fragmentation. All the frontiers and divisions that we make originate in thought. (Bohm 1996)

Art, science, and technology, although inseparable, organize institutionally in ways that legitimize their separate activities. The energy (e.g., extraction of fossil fuel), medico-pharmaceutical, and military industries fund scientific research, while art is underwritten by organizations that grant and market packages for tourists and others who can afford leisurely entertainment. Economic interests that marginalize all but market priorities tend to control and limit creativity.

Trespassing as Connection and Recombination

Agitation and uncertainty marked social experience in the twentieth century. Reactive advocacy of non-linearity in art and science confronted mechanistic, deterministic, pre-programmed concepts. In Zürich, Berlin, Paris, and New York, Dadaists questioned established morals, values, and norms. They challenged the bourgeois collective imagination that they perceived as anchored in a static, predestined world. Transformation through action began to define new practices. Through performances, installations, and other productions, through body art, and through "happenings," human bodies themselves acted as catalysts of social awareness. The transition from a world of "art objects" to one of "processes" occurred toward the end of the century and correlated with changes in science. Gaia Theory (Lovelock 1972; Harding and Margulis 2009; see chapters 8 and 10) conceives of a joint atmosphere, hydrosphere, and lithosphere on Earth's surface that includes a biosphere as the place where life exists. Gaia recognizes the incessant flow of matter and energy that connects these "spheres." "The Earth is indivisible, wrote McHarg (2006). The atmosphere and surface sediments are part of Gaia, a live planetary physiological system of material and energetic transformation. Lovelock posited life and environmental interaction as a "colligative" characteristic of Earth. Continuous adjustments are made by all organisms in ways that help modulate temperature, chemical composition, and structure at Earth's surface. Environment and life together—a "tightly coupled system," in Lovelock's words—make a continuum.

Assertions that atmospheric processes interact with life and form a "planetary circulatory system" parallel vanguard artistic activities isolated from mainstream culture. Dadaists and Futurists, who emphasized transgression, connection, and recombination, attempted to transform rigid walls of the "museum" into sensitive membranes permeable to the outside. Bauhaus and Constructivism linked fine art to its applications in design, architecture, urban renewal, and education. László Moholy-Nagy and his contemporaries insisted that creations are useful only when they produce previously unknown relationships.

Gaia

Gaia theory changes our perception of organic vs. inorganic and animate vs. inanimate matter (Harding 2009). It reduces these dichotomies and stimulates research in distinct disciplines by reconnecting Earth sciences with life sciences. "Natural history" is returning under a new name. Innovative artistic practices that consolidate dynamic and productive threads of twentieth-century aesthetics currently attempt to overrule the banal assumption that art and life are dichotomous. The simultaneous transgression of limits in both art and science was emphasized by Félix Guattari, who aspired to establish an "ecosophy" that would link the environmental science of ecology to "social and mental ecology" (Guattari 1992). That art and science manifest discovery and creativity was, Guattari claimed, "the outstanding expression of a feature common to all levels of nature" (Prigogine 1997). By the 1990s, exhibition centers, research departments, and festivals had erupted around the world to ameliorate institutionalized crises. Critical of social change that stems from scientific research and technological innovation, international multicultural activities (e.g., seminars, forums, workshops, exhibits in science museums and in children's museums) helped hybridize knowledge.

Work is dynamic. Time is programmable. Much broadcast artistic achievement is ephemeral and intangible. On video, on film, in multimedia, in interactive media, and even in commercial advertising, artists formulate fluid, transformable worlds. They generate chimeric evolutionary entities. Different realities are connected, ranging from the nanometric scales of bioengineering to global dynamics that extend hundreds of kilometers.

The visionaries Vladimir Ivanovich Vernadsky and Pierre Teilhard de Chardin recognized humans as a new stage in biogeochemical evolution, a "thinking layer" of organized matter that accelerates growth and change at Earth's surface. Technology is an integral, accelerating part of the planetary biosphere (Grinevald 1998). Dorion Sagan (2007) writes:

Thought, like life itself, is a flow of matter and energy; the body is its complement. Thinking and being are aspects of the same physical organization and its actions. If we accept the basic continuity between body and mind, thought is, in essence, like the rest of physiology and behavior. Thinking, like excretion and eating, is the result of interactions in the organism's chemistry. . . . If what is called "thought" was caused by this type of interaction, perhaps communication among organisms, each of them a thinker, might lead to a larger process than individual thought. This may be implicit in Vernadsky's notion of the nöosphere.

From Neurons to Networked Societies

Barely 100 years separate Ramón y Cajal's discovery of neuronal networks from Manuel Castells's theories (1996) about the Network Society. Transformation of social organization accelerates. Today's citizens are incorporated into dynamic, changing, petroleum-dependent systems. Energy, matter, and knowledge move incessantly. Activities of others are sensed and responded to by members of 30 million different species. Exploration of our world should become an adventure.[1] The flowing

1. Banquete (www.banquete.org), a research project that I have developed over 15 years with Karin Ohlenschläger, explores the relations among biological, socioeconomic, and informational processes. Since the early 1990s, it has brought outstanding creators and researchers (anthropologists, artists, architects, biologists, philosophers, economists, neuroscientists, sociologists) together at various events (expositions, workshops, symposia, seminars, publications) to reflect on the patterns and processes of transformation that govern tangible and intangible flows of matter, energy, and information. Banquete arose from a challenge: for many people to devise and build an interaction space [ACTS (Art-Science-Technology-Society)] to investigate new forms of knowledge creation and transfer, i.e., a context for practicing a transversal way of thinking that explores the interaction among ecosystems, the technosphere, and the symbolic universes that define the individual and collective imagination of contemporary societies.

space that defines the globalized world of the present century yields to a new structure with multiple nodes and networks. A continuous flow of information and dialog within a network occurs where each point is a node, an outlook, and a story with a function. A new system of poly-centered, dynamic and horizontal productions and their transmission appear. Ideas, material entities, and even institutions are catalysts for the process of cultural transformation, whose behavior is more akin to physiology than to mechanics.

In this so-called Age of Information and Knowledge, with biological, electronic, and cognitive packages all parts of the system, can cultural institutions transform to reflect the learning flexibility that cellular and neuronal sensory systems evolved since the opening of the Phanerozoic eon (542 million years ago or earlier)? Can "eco-physiological" genera-tion and dissemination of knowledge prevail? Not unless the mechanical, deterministic, and compartmentalized belief systems that impede cultural flow are abandoned and replaced with more holistic views (Harding 2009). Reconnections and reconfigurations that in the past were sepa-rated by anthropocentric, linear, and dichotomous habits of thought may help return us to nature. New integrated art, science, and technology will usher in natural connections between twentieth-century Gaian and sym-biogenetic ideas (see chapters 8 and 13) that lead to worldwide cultural patterns worthy of nurture in the current century.

Cosmologists and astronomers tell us that only 3 percent of the uni-verse is detected by us. The remaining 97 percent is "dark matter and energy." The magnitude of uncertainty to be navigated is immense. Implementation of what we can hardly even imagine lies ahead. In spite of our limited, ancient sensory systems, our anthropocentrism, prejudi-cial symbolism, biased educations, dangerous ignorance, and distracting technologies, our imaginations tend to soar. We can only hope that the interconnectedness of creative activities that has led to transformation of the human environment will continue to generate surprise. We should only act to nurture healthy humans in numbers that can be supported by the local environments of each. We anticipate artistic-scientific infor-matic concepts and tools to implement sustained innovation that includes us humans in the rest of nature, from which we came and to which we will always belong.

Bibliography

Adler, J. 1988. Chemotaxis: Old and new. *Botanica Acta* 101: 93–10.

Aitken, R. 1984. *The Mind of the Clover: Essays in Zen Buddhist Ethics*. North Point.

Alexandre, G., S. Greer-Phillips, and I. B. Zhulin. 2004. Ecological role of energy taxis in microorganisms. *FEMS Microbiology Reviews* 28: 113–126.

Alliegro, M. C., and M. A. Alliegro. 2008. Centrosomal RNA correlates with intron-poor nuclear genes in *Spisula* oocytes. *Proceedings of the National Academy of Sciences* 105: 6993–6997.

Alliegro, M. A., J. J. Henry, and M. C. Alliegro. 2010. Rediscovery of the nucleolinus, a dynamic organelle. *Proceedings of the National Academy of Sciences*, in press.

Anderson, D. L. 1992. The Earth's interior. In *Understanding the Earth*, ed. G. Brown, C. Hawkesworth, and R. Wilson. Cambridge University Press.

Aravind, L., and E. V. Koonin. 1999. DNA polymerase-like nucleotidyltransferase superfamily: identification of three new families, classification and evolutionary history. *Nucleic Acids Research* 27: 1609–1618.

Armitage, J. P. 1999. Bacterial tactic responses. *Advances in Microbial Physiology* 41: 229–289.

Asikainen, C. A., and S. Werle. 2007. Fresh water iron-manganese microbialites from the Connecticut River headwaters. *Proceedings of the National Academy of Sciences* 104: 17597–17581.

Balfour, F. M. 1880–1881. *A Treatise on Comparative Embryology*. Macmillan.

Bateson, G. 2002. *Mind and Nature: A Necessary Unity*. Hampton.

Bearder, S. K. 1987. Lorises, bushbabies and tarsiers: Diverse societies in solitary foragers. In *Primate Societies*, ed. B. Smuts, D. Cheney, R. Seyfarth, R. Wrangham, and T. Struhsaker. University of Chicago Press.

Beauchamp, G. K., R. L. Doty, D. G. Moulton, and R. A. Mugford. 1976. The pheromone concept in mammalian chemical communication: A critique. In *Mammalian Olfaction, Reproductive Processes, and Behavior*, ed. R. Doty. Academic Press.

Becerra, A., L. Delay, A. Islas, and A. Lazcano. 2007. Very early stages of biological evolution related to the nature of the last common ancestor of the three major cell domains. *Annual Review of Ecology Evolution and Systematics* 38: 361–379.

Beliaev, A. S., D. K. Thompson, M. W. Fields, L. Y. Wu, D. P. Lies, K. H. Nealson, and J. Z. Zhou. 2002. Microarray transcription profiling of a *Shewanella oneidensis* etrA mutant. *Journal of Bacteriology* 184: 4612–4616.

Ben-Jacob, E. 2003. Bacterial self-organization: Co-enhancement of complexification and adaptability in a dynamic environment. *Philosophical Transactions of the Royal Society* A361: 1283–1312.

Ben-Jacob, E., I. Becker, Y. Shapira, and H. Levine. 2004. Bacterial linguistic communication and social intelligence. *Trends in Microbiology* 12: 366–372.

Ben-Jacob, E., Y. Shapira, and A. I. Tauber. 2006. Seeking the foundations of cognition in bacteria: From Schrödinger's negative entropy to latent information. *Physica A* 359: 495–524.

Bijma, J., J. Erez, and C. Hemleben. 1990. Lunar and semi-lunar reproductive cycles in some spinose planktonic foraminifers. *Journal of Foraminiferal Research* 20: 117–127.

Blumberg, M. 2009. *Freaks of Nature: What Anomalies Tell Us About Development and Evolution*. Oxford University Press.

Bohm, D. 1996. *On Dialogue*. Routledge.

Bohm, D. 2002. *Wholeness and the Implicate Order*. Routledge.

Bonner, J. T. 1998. A way of following individual cells in the migrating slugs of *Dicyostelium discoideum*. *Proceedings of the National Academy of Sciences* 95: 9355–9359.

Bortz, W. 1984. The disuse syndrome. *Western Journal of Medicine* 141: 69–98.

Bortz, W. 1985. Physical exercise as an evolutionary force. *Journal of Human Evolution* 14: 145–156.

Bowen, M. 1978. *Family Therapy in Clinical Practice*. Jason Aronson.

Boyle, P. J., and R. Mitchell. 1978. Absence of microorganisms in crustacean digestive tracts. *Science* 200: 1157–1159.

Brehm, U., W. E. Krumbein, and K. A. Palinska. 2006. Biomicrospheres generate ooids in the laboratory. *Geomicrobiology Journal* 23: 545–550.

Broderick, N. A., K. F. Raffa, R. M. Goodman, and J. Handelsman. 2004. Census of the bacterial community of the gypsy moth larval midgut by using culturing and culture-independent methods. *Applied and Environmental Microbiology* 70: 293–300.

Brown, F. A., J. W. Hastings, and J. D. Palmer. 1970. *The Biological Clock: Two Views*. Academic Press.

Brown, S., L. Margulis, L. Ibarra, and D. Siqueiros. 1985. Desiccation resistance contamination as mechanisms of Gaia. *BioSystems* 17: 337–360.

Bucher, W. H. 1933. *Deformation of the Earth's Crust*. Princeton University Press.

Byrne, R. 1995. *The Thinking Ape: Evolutionary Origins of Intelligence*. Oxford University Press.

Castells, M. 1996. *The Rise of the Network Society*. Blackwell.

Cavalier-Smith, T. 2003. Microbial muddles. *BioScience* 53: 1008–1010.

Chapman, M. J., and M. C. Alliegro. 2007. A symbiogenetic basis for the centrosome? *Symbiosis* 44: 23–32.

Chapman, M. F., M. F. Dolan, and L. Margulis. 2000. Centrioles and kinetosomes: Form, function and evolution. *Quarterly Review of Biology* 75: 409–429.

Charles-Dominique, P. 1977. *Ecology and Behaviour of Nocturnal Primates: Prosimians of Equatorial West Africa*. Duckworth.

Charlson, R. J., J. E. Lovelock, M. O. Andreae, and S. G. Warren. 1987. Oceanic phytoplankton, atmospheric sulphur, cloud albedo and climate. *Nature* 326: 655–661.

Cockell, C. S. 2001. A photobiological history of Earth. In *Ecosystems, Evolution, and Ultraviolet Radiation*, ed. C. Cockell and A. Blaustein. Springer-Verlag.

Connor, R. C., A. J. Read, and R. Wrangham. 2000. The Bottlenose Dolphin: Social Relationships in a Fission-Fusion Society. In *Cetacean Societies: Field Studies of Dolphins and Whales*, ed. J. Mann, R. Connor, P. Tyak, and H. Whitehead. University of Chicago Press.

Crewe, R. M., B. V. Burger, M. Le Roux, and Z. Katsir. 1979. Chemical constituents of the chest gland secretion of the thick-tailed galago *Galago crassicaudatus*. *Journal of Chemical Ecology* 5: 861–868.

Crist, E., and H. B. Rinker. 2009. *Gaia in Turmoil: Climate Change, Biodepletion, and Earth Ethics in An Age of Crisis*. MIT Press.

Curley, E. 1994. *A Spinoza Reader: The Ethics and Other Works by Benedict De Spinoza*. Princeton University Press.

Danchin, A. 1993. Phylogeny of adenlyl cyclases. *Advances in Second Messenger and Phosphoprotein Research* 27: 109–162.

Darwin, C. 1859. *On the Origin of Species by Means of Natural Selection, or The Preservation of Favoured Races in the Struggle for Life*, first edition. John Murray.

Darwin, C. R. 1868. *The Variation of Animals and Plants Under Domestication*, first edition, volume 2. John Murray.

Darwin, C. 1883. *The Variation of Animals and Plants under Domestication*, second edition. Appleton.

Darwin, C. 1898 [1965]. *The Expression of the Emotions in Man and Animals*. University of Chicago Press.

Dawkins, R. 1990. *The Selfish Gene*, second edition. Oxford University Press.

De Groot, N., N. Otting, G. G. M. Doxiadis, S. S. Balla-Jhagjhoorsingh, J. L. Heeney, J. J. van Rood, P. Gagneux, and R. E. Bontrop. 2002. Evidence for an ancient selective sweep in the MHC class I gene repertoire of chimpanzees. *Proceedings of the National Academy of Sciences* 99: 11748–11753.

Del Pero, M., L. Pozzi, and J. C. Masters. 2006. A composite molecular phylogeny of living Lemuroid primates. *Folia Primatologica* 77: 434–445.

Denny, M. 1993. *Air and Water: The Biology and Physics of Life's Media.* Princeton University Press.

Derrida, J. 1982. *Positions.* University of Chicago Press.

De Waal, F. 1982. *Chimpanzee Politics: Power and Sex Among Apes.* Harper and Row.

Dick, P. K. 1953. Impostor. *Astounding Science Fiction*, June.

Dick, P. K. 1968. *Do Androids Dream of Electric Sheep?* Ballantine Books.

Dolan, M. 2005. The missing piece. In *Microbial Phylogeny and Evolution: Concepts and Controversies*, ed. J. Sapp. Oxford University Press.

Dolan, M., H. Melnitsky, L. Margulis, and R. Kolnicki. 2004. Motility proteins and the origin of the nucleus. *Anatomical Record* 268: 290–301.

Doolittle, W. F. 1981. Is nature really motherly? *Coevolution Quarterly* 29: 58–63.

Dubinina, G. A., M. Y. Grabovich, N. Leshcheva, F. A. Rainey, and E. Y. Gavrish. 2010. *Spirochaeta perfilievii* sp. nov., oxygen-tolerant, sulfide oxidizing, sulfur and thiosulfate-reducing spirochete isolated from a saline spring. *International Journal of Systematic and Evolutionary Microbiology*, in press.

Dunny, G. M., and S. C. Winans, eds. 1999. *Cell-Cell Signaling in Bacteria.* American Society of Microbiology Press.

Einstein, A., and M. Born. 2004. *The Born-Einstein Letters: Friendship, Politics and Physics in Uncertain Times.* Macmillan.

Eldredge, N., and S. J. Gould. 1972. Punctuated equilibria: An alternative to phyletic gradualism. In *Models in Paleobiology*, ed. T. Schopf. Freeman-Cooper.

Endler, J. A. 1992. Signals, signal conditions, and the direction of evolution. *American Naturalist* 139: s125–s153.

Endres, K., and W. Schad. 1997. *Moon Rhythms in Nature.* Floris.

Engebrecht, J., K. H. Nealson, and M. Silverman. 1983. Bacterial bioluminescence: isolation and genetic analysis of the functions from *Vibrio fischeri. Cell* 32: 773–781.

Fleck, L. 1935 [1979]. *Genesis and Development of a Scientific Fact.* University of Chicago Press.

Folch, R., ed. 2000. *Encyclopedia of the Biosphere: A Guide to the World's Ecosystems.* Gale Group.

Foster, R. G., and L. Kreitzman. 2004. *Rhythms of Life: The Biological Clocks That Control the Daily Lives of Every Living Thing.* Yale University Press.

Fry, I. 2000. *The Emergence of Life.* Rutgers University Press.

Gagneux, P., C. Wills, U. Gerloff, D. Tautz, P. A. Morin, C. Boesch, B. Fruth, G. Hohmann, O. A. Ryder, and D. S. Woodruff. 1999. Mitochondrial sequences show diverse evolutionary histories of African hominoids. *Proceedings of the National Academy of Sciences* 96: 5077–5082.

Galperin, M. Y., and M. Gomelsky. 2005. Bacterial signal transduction modules: from genomics to biology. *American Society of Microbiology News* 71: 326–333.

Garland, C. D., G. V. Nash, and T. A. McMeekin. 1982. Absence of surface-associated microorganisms in adult oysters Crassostrea gigas. *Applied and Environmental Microbiology* 44: 1205–1211.

Glanz, J. 1997. Force-carrying web pervades living cell. *Science* 276: 678–679.

Godfrey, L. R., and J. C. Masters. 2000. Comments on Kolnicki: Kinetochore reproduction theory may explain rapid chromosome evolution. *Proceedings of the National Academy of Sciences* 97: 9821–9823.

Grinevald, J. 1998. Introduction: The invisibility of the Vernadskian revolution. In *The Biosphere*, ed. V. Vernadsky. Copernicus/Springer-Verlag.

Grishanin, R. N., I. I. Chalmina, and I. B. Zhulin. 1991. Behavior of Azospirillum brasiliense in a spatial gradient of oxygen and in a redox gradient of an artificial electron-acceptor. *Journal of General Microbiology* 137: 2781–2785.

Guattari, F. 1992. *Pour une refondation des practiques sociales.* Le Monde Diplomatique.

Guisto, J. P., and L. Margulis. 1981. Karyotypic fission theory and the evolution of Old World monkeys and apes. *BioSystems* 13: 267–302.

Haeckel, E. 1866. *Generelle Morphologie der Organismen.: Allgemeine Grundzüge der organischen Formen-Wissenschaft, mechanisch begründet durch die von Charles Darwin reformirte Descencendenz-Theorie.* Georg Reimer.

Hall, J. L. 2011. Spirochete contribution to the eukaryotic genome. *Proceedings of the National Academy of Sciences*, submitted.

Harding, S. 2009. *Animate Earth: Science, Intuition and Gaia*, second edition. Green Books.

Harding, S., and L. Margulis. 2009. Water Gaia: Three and a half thousand million-years of wetness on planet Earth. In *Gaia in Turmoil: Climate Change, Biodepletion, and Earth Ethics in An Age of Crisis*, ed. E. Crist and H. Rinker. MIT Press.

Hayden, D. 2003. *Pox: Genius, Madness, and the Mysteries of Syphilis.* Basic Books.

Hazen, R. M. 2005. *Genesis: The Scientific Quest for Life's Origins.* National Academies Press.

Henke, J. M., and B. L. Bassler. 2004. Three parallel quorum sensing systems regulate gene expression in *Vibrio harveyi*. *Journal of Bacteriology* 186: 6902–6914.

Herndon, J. M. 2010. The natural nuclear reactor at the core of the Earth. *American Scientist*, submitted.

Hoffman, P. F., and D. P. Schrag. 2000. Snowball Earth. *Scientific American* 282: 68–75.

Holmes, M. G., ed. 1991. *Photoreceptor Evolution and Function*. Academic Press.

Imai, H. T., T. Maruyama, T. Gojobori, Y. Inoue, and R. H. Crozier. 1986. Theoretical bases for karyotype evolution. The minimum interaction hypotheses. *American Naturalist* 128: 900–920.

Jerison, H. J. 1997. Evolution of prefrontal cortex. In *Development of the Prefrontal Cortex: Evolution, Neurobiology, and Behavior*, ed. N. Krasnegor, G. Lyon, and P. Goldman-Rakic. Brookes.

Jonas, H. 1979. *The Phenomenon of Life: Toward A Philosophical Biology*. Greenwood.

Joyce, G. F. 2002. The antiquity of RNA-based evolution. *Nature* 418: 214–221.

Kahle, W., and M. Frotscher. 2005. *Taschenatlas Anatomie, Nervensystem und Sinnesorgane*. vol. 3. Thieme.

Khakhina, L. N. 1992. *Concepts of Symbiogenesis: A Historical and Critical Study of the Research of Russian Botanists*. Yale University Press.

Kolnicki, R. 2000. Kinetochore reproduction in animal evolution: Cell biological explanation of karyotypic fission theory. *Proceedings of the National Academy of Sciences* 97: 9493–9497.

Kozo-Polyansky, B. M. 1924 [2010]. *Symbiogenesis: A New Principle of Evolution*. Harvard University Press.

Kremer, H. 2008. *The Silent Revolution in Cancer and AIDS Medicine*. Xlibris.

Krumbein, W. E. 2008. Biogenerated rock structures. *Space Science Reviews* 135: 81–94.

Krumbein, W. E., U. Brehm, G. Gerdes, A. A. Gorbushina, G. Levit, and K. A. Palinska. 2003. Biofilm, biodictyon, and biomat—Biolaminites, oolites, stromatolites—Geophysiology, global mechanisms and parahistology. In *Fossil and Recent Biofilms: A Natural History of Life on Earth*, ed. W. Krumbein, D. Paterson, and G. Zavarzin. Kluwer.

Kumar, R. K., and M. Yarus. 2001. RNA-catalyzed amino acid activation. *Biochemistry* 40: 6998–7004.

Kung, C. 2005. A possible unifying principle for mechanosensation. *Nature* 436: 647–654.

Lapo, A. 1987. *Traces of Bygone Biospheres*. Mir.

Lean, J. 1997. The Sun's variable radiation and its relevance for Earth. *Annual Review of Astronomy and Astrophysics* 35: 33–67.

Leff, H. S., and A. F. Rex. 1990. *Maxwell's Demon*. Princeton University Press.

Leibniz, G. W. 1714 [2008]. *The Monadology*. Forgotten Books.

Leidy, J. 1850. On the existence of endophyta in healthy animals as a natural condition. *Proceedings. Academy of Natural Sciences of Philadelphia* 4: 225–229.

Lettvin, J. 1973. Polaroid Interactive Lecture.

Ley, R. E., P. J. Turnbaugh, S. Klein, and J. I. Gordon. 2006. Human gut microbes associated with obesity. *Nature* 444: 1022–1023.

Li, X., and R. B. Nicklas. 1995. Mitotic forces control a cell-cycle checkpoint. *Nature* 373: 630–632.

Lovelock, J. E. 1972. Gaia as seen through the atmosphere. *Atmospheric Environment* 6: 579–580.

Lovelock, J. E. 1988. *The Ages of Gaia: A Biography of Our Living Earth*. Norton.

Lovelock, J. E. 1991. *Gaia: The Practical Science of Planetary Medicine*. Gaia Books, 130–131.

Lovelock, J. E. 2006. *The Revenge of Gaia: Earth's Climate in Crisis and the Fate of Humanity*. Penguin.

Lowman, P. D., Jr. 1976. Crustal evolution in silicate planets: Implications for the origin of continents. *Journal of Geology* 84: 1–26.

Lowman, P. D., Jr. 2002. *Exploring Space, Exploring Earth*. Cambridge University Press.

MacBride, E. W. 1914. *Text-Book of Embryology*, volume 1: *Invertebrata*. Macmillan.

Maniotis, A., K. Bojanowski, and D. Ingber. 1997. Mechanical continuity and reversible chromosome disassembly within intact genomes removed from living cells. *Journal of Cellular Biochemistry* 65: 114–130.

Manuelidis, L. 1990. A view of interphase chromosomes. *Science* 250: 1533–1540.

Margulis, L. 1993. *Symbiosis in Cell Evolution: Microbial Communities in the Archean and Proterozoic Eons*, second edition. Freeman.

Margulis, L. 1998. *Symbiotic Planet: A New Look at Evolution*. Basic Books.

Margulis, L. 2004. Serial endosymbiotic theory SET and Composite individuality: Transition from bacterial to eukaryote genomes. *Microbiology Today* 31: 172–174.

Margulis, L. 2005. Jointed threads. *Natural History*, June: 28–32.

Margulis, L. 2007. Life as Growth, Lynn Margulis's interview with William Day. In *Mind, Life and Universe: Conversations with Great Scientists of Our Time*, ed. L. Margulis and E. Punset. Chelsea Green.

Margulis, L., and M. J. Chapman. 2010. *Kingdoms and Domains: An Illustrated Guide to the Phyla of Life on Earth*, fourth edition. Academic Press.

Margulis, L., and M. J. Chapman, eds. 2012. *Handbook of Protoctista: The Structure, Cultivation, Habitats and Life Histories of the Eukaryotic Microorganisms and Their Descendants Exclusive of Animals, Plants and Fungi*, second edition. Jones and Bartlett.

Margulis, L., M. Chapman, R. Guerrero, and J. Hall. 2006. The last eukaryotic common ancestor LECA: Acquisition of cytoskeletal motility from aerotolerant spirochetes in the Proterozoic Eon. *Proceedings of the National Academy of Sciences* 103: 13080–13085.

Margulis, L., M. Dolan, and J. Whiteside. 2005. Imperfections and oddities in the origin of the nucleus, ed. E. Verba and N. Eldredge. *Paleobiology* 31: 175–191.

Margulis, L., and R. Fester, eds. 1991. *Symbiosis As A Source of Evolutionary Innovation: Speciation and Morphogenesis*. MIT Press.

Margulis, L., and J. E. Lovelock. 1974. Biological modulation of the Earth's atmosphere. *Icarus* 21: 474–489.

Margulis, L., C. Matthews, and A. Haselton, eds. 2000. *Environmental Evolution: Effects of the Origin and Evolution of Life on Planet Earth*, second edition. MIT Press.

Margulis, L., and L. Olendzenski, eds. 1992. *Environmental Evolution: The Effect of the Origin and Evolution of Life on Planet Earth*. MIT Press.

Margulis, L., and D. Sagan. 2007. *Dazzle Gradually: Reflections on the Nature of Nation*. Chelsea Green.

Margulis, L., and D. Sagan. 2002. *Acquiring Genomes: A Theory of the Origins of Species*. Basic Books.

Masters, J. C. 1991. Loud calls of *Galago crassicaudatus* and *G. garnettii* and their relation to habitat structure. *Primates* 32: 153–167.

Masters, J. C. 1993. *Primates and Paradigms: Problems with the Identification of Genetic Species*, ed. W. Kimbel and L. Martin. Plenum.

Mayer, J. R. 1845. *Die organische Bewegung in ihrem Zusammenhange mit dem Stoffwechsel*. Drechsler'sche Buchhandlung.

McClendon, A. 1901. Chromosome isolation from *Cerabratulus* eggs. William Roux Archives, Woods Hole, Massachusetts.

McHarg, I. 2006. *Ian McHarg: Conversations with Students: Dwelling in Nature*, ed. L. Margulis, J. Corner, and B. Hawthorne. Princeton Architectural Press.

McKeon, R., ed. 2001. *The Basic Works of Aristotle*. Princeton University Press.

Melnitsky, H., and L. Margulis. 2004. Centrosomal proteins in termite symbionts: Gamma-tubulin and scleroderma antibodies bind rotation zone of *Caduceia versatilis*. *Symbiosis* 37: 323–333.

Miller, S. L. 1953. Production of amino acids under possible primitive Earth conditions. *Science* 117: 528–529.

Miller, S. L., and H. C. Urey. 1959. Organic compound synthesis on the primitive Earth: Several questions about the origin of life have been answered, but much remains to be studied. *Science* 130: 245–251.

Minkel, J. R. 2002. The exergy machine: Have we finally found life's true purpose? *New Scientist* 40: 200–300.

Moalem, S., and J. Prince. 2007. *Survival of the Sickest: A Medical Maverick Discovers Why We Need Disease.* William Morrow.

Moore, W. S., W. E. Dean, S. Krishnaswami, and D. V. Borole. 1980. Growth rates of manganese nodules in Oneida Lake, New York. *Earth and Planetary Science Letters* 46: 191–200.

Moore-Ede, M. C., F. M. Sulzman, and C. A. Fuller. 1982. *The Clocks That Time Us: Physiology of the Circadian Timing System.* Harvard University Press.

Moran, N. A., P. H. Degnan. S. R., Santos, H. E. Dunbar, and H. Ochman, H. 2005. The players in a mutualistic symbiosis: Insects, bacteria, viruses, and virulence genes. *Proceedings of the National Academy of Sciences* 102: 16919–16926.

Morowitz, H., J. D. Kostelnik, J. Yang, and G. D. Cody. 2000. The origin of intermediary metabolism. *Proceedings of the National Academy of Sciences* 97: 7704–7708.

Myers, C. R., and K. H. Nealson. 1988. Bacterial manganese reduction and growth with manganese oxide as the sole electron acceptor. *Science* 240: 1319–1321.

Nagele, R. G., A. Q. Velasco, W. J. Anderson, D. J. McMahon, Z. Thomson, and J. Fazekas. 2001. Maintenance of interphase chromosome topology. *Journal of Cell Science* 114: 377–388.

Nealson, K. H., D. P. Moser, and D. A. Saffarini. 1995. Anaerobic electron acceptor chemotaxes in *Shewanella putrefaciens. Applied and Environmental Microbiology* 61: 1551–1554.

Nevo, E. 1999. *Mosaic Evolution of Subterranean Mammals: Regression, Progression, and Global Gonvergence.* Oxford University Press.

Nietzsche, F. 2003. *The Genealogy of Morals.* Dover.

Nisbet, E. G. 1987. *The Young Earth.* Allen and Unwin.

Okamoto, N., and I. Inouye. 2006. *Hatena arenicola* gen. et ap. nov., a Katablepharid undergoing probable plastid acquisition. *Protist* 157: 401–419.

O'Neill, R. J. W., M. J. O'Neill, and J. A. M. Graves. 1998. Undermethylation associated with retroelement activation and chromosome remodelling in an interspecific mammalian hybrid. *Nature* 393: 68–72.

Palmer, J. D. 1995. *Biological Rhythms and Clocks of Intertidal Animals.* Oxford University Press.

Pariente, G. F. 1975. Observation opthalmologique de zones fovéales vraies chez Lemur catta et *Hapalemur griseus*, primates de Madagascar. *Mammalia* 39: 487–497.

Pastorini, J., U. Thalmann, and R. D. Martin. 2003. A molecular approach to comparative phylogeny of extant Malagasy lemurs. *Proceedings of the National Academy of Sciences* 100: 5879–5884.

Pearce, J. M. 1997. *Animal Learning and Cognition*. Psychology Press.

Perry, J., H. Slater, and K. Choo. 2004. Centric fission—simple and complex mechanisms. *Chromosome Research* 12: 627–640.

Petter, J. J., and P. Charles-Dominique. 1979. Vocal communication in prosimians. In *The Study of Prosimian Behavior*, ed. G. Doyle and R. Martin. Academic Press.

Popa, R., P. K. Weber, J. Pett-Ridge, J. A. Finzi, S. J. Fallon, I. D. Hutcheon, K. H. Nealson, and D. G. Capone. 2007. Carbon and nitrogen fixation and metabolite exchange in and between individual cells of *Anabaena oscillarioides*. *International Society of Microbial Ecology* 1: 354–360.

Prigogine, I. 1997. *End of Certainty*. Free Press.

Rees, D. C., and J. B. Howard. 2003. The interface between the biological and inorganic worlds: iron-sulfur metalloclusters. *Science* 300: 929–931.

Roth, G. 2000. The evolution of consciousness. In *Brain, Evolution and Cognition*, ed. G. Roth and M. Wullimann. Wiley-Spektrum.

Roth, G., and U. Dicke. 2005. Evolution of the brain and intelligence. *Trends in Cognitive Sciences* 9: 250–257.

Rumpho, M. E., J. M. Worful, J. Lee, K. Kannana, M. S. Tyler, D. Bhattachary, A. Moustaf, and J. R. Manhart. 2008. Horizontal gene transfer of the algal nuclear gene psbO to the photosynthetic sea slug *Elysia chlorotica*. *Proceedings of the National Academy of Sciences* 105: 17867–17871.

Ryan, F. P. 2007. Viruses as symbionts. *Symbiosis* 44: 11–21.

Sagan, D. 2007. *Notes from the Holocene: A Brief History of the Future*. Chelsea Green.

Sagan, D., and J. Whiteside. 2004. Gradient-reduction theory: Thermodynamics and the purpose of life. In *Scientists Debate Gaia: A Next Century*, ed. H. Schneider, J. Miller, E. Crist, and P. Boston. MIT Press.

Sapp, J. 1994. *Evolution by Association: A History of Symbiosis*. Oxford University Press.

Sapp, J. 2005. The prokaryote-eukaryote dichotomy: meanings and mythology. *Microbiology and Molecular Biology Reviews* 69: 292–305.

Sapp, J. 2009. *The New Foundations of Evolution: On the Tree of Life*. Oxford University Press.

Sapp, J., F. Carrapico, and M. Zolotonosov. 2002. Symbiogenesis: The hidden fact of Constantin Mereschkowsky. *History and Philosophy of the Life Sciences* 24: 413–440.

Schneider, E. D., and J. J. Kay. 1989. Nature abhors a gradient. In *Proceedings of the 33rd Annual Meeting of the International Society for the Systems Sciences*.

Schneider, E. D., and D. Sagan. 2005. *Into the Cool: Energy Flow, Thermodynamics and Life*. University of Chicago Press.

Schoenheimer, R. 1942. *The Dynamic State of Body Constituents*. Harvard University Press.

Schopf, J. W. 2006. Fossil evidence of Archaean life. *Philosophical Transactions of the Royal Society of London. Series B, Biological Sciences* 361: 869–885.

Schrödinger, E. 1994. *What Is Life? The Physical Aspect of the Living Cell*. Cambridge University Press.

Schwartzman, D. 1999. *Life, Temperature, and the Earth: The Self-Organizing Biosphere*. Columbia University Press.

Scofield, B. 2004. Precedents to Gaia. In *Scientists Debate Gaia: The Next Century*, ed. S. Schneider, et al. MIT Press.

Scofield, B., and L. Margulis. 2011. Psychological discontent: Self and science on our symbiotic planet. In *Ecopsychology in a Technological World*, ed. P. Hasbach et al. MIT Press.

Shapiro, J., and M. Dworkin, eds. 1997. *Bacteria As Multicellular Organisms*. Oxford University Press.

Shenoy, A. R., and S. S. Visweswariah. 2004. Class III nucleotide cyclases in bacteria and archaebacteria: Lineage-specific expansion of adenylyl cyclases and a dearth of guanylyl cyclases. *FEBS Letters* 561: 11–21.

Skoyles, J. 2008a. Human metabolic adaptations and prolonged expensive neurodevelopment: A review. Available at http: //precedings.nature.com

Skoyles, J. 2008b. Respiratory, postural and spatio-kinetic motor stabilization, internal models, top-down timed motor coordination and expanded cerebello-cerebral circuitry: A review. Available at http: //precedings.nature.com.

Skoyles, J. R., and D. Sagan. 2002. *Up from Dragons: The Evolution of Intelligence*. McGraw-Hill

Sly, B. J., M. S. Snoke, and R. A. Raff. 2003. Who came first—larvae or adults? Origins of bilaterian metazoan larvae. *International Journal of Developmental Biology* 47: 623–632.

Sontag, S. 2001. *Illness as Metaphor and AIDS and Its Metaphors*. Picador.

Sullivan, J. T., and C. W. Ronson. 1998. Evolution of rhizobia by acquisition of the 500-kb symbiosis island that integrates into a phe-tRNA gene. *Proceedings of the National Academy of Sciences* 5: 5145–5149.

Sweeney, B. M. 1987. *Rhythmic Phenomena in Plants*. Academic Press.

Tattersall, I. 2006. The concept of cathemerality: History and definition. *Folia Primatologica* 77: 7–14.

Taylor, B., J. B. Miller, H. M. Warrick, and D. E. Koshland, Jr. 1979. Electron acceptor taxis and blue light effect on bacterial chemotaxis. *Journal of Bacteriology* 140: 567–573.

Todd, N. 2000. Mammalian evolution: Karyotypic fission theory. In *Environmental Evolution: Effects of the Origin and Evolution of Life on Planet Earth*, ed. L. Margulis, C. Matthews, and A. Haselton, second edition. MIT Press.

Turner, J. S. 2000. *The extended organism: The physiology of animal-built structures*. Cambridge: Harvard University Press.

Turner, J. S. 2007. *The Tinkerer's Accomplice: How Design Emerges from Life Itself*. Harvard University Press.

Vernadsky, V. I. 1998. *The Biosphere*. Nevraumont/Springer-Verlag.

Vernadsky, V. I. 1945. The biosphere and the nöosphere. *American Scientist* 33: 1–12.

Villarreal, L. P. 2007. Virus-host symbiosis mediated by persistence. *Symbiosis* 44: 1–9.

Vinnikov, Y. A. 1982. *Evolution of Receptor Cells, Molecular Biology Biochemistry and Biophysics*. Springer-Verlag.

Wallin, I. E. 1927. *Symbionticism and the Origin of Species*. Wiley.

Waser, P. M., and C. H. Brown. 1986. Habitat acoustics and primate communication. *American Journal of Primatology* 10: 135–154.

Wegener, A. 1924. *The Origin of Continents and Oceans*. Dutton.

Westbroek, P. 1991. *Life as a Geological Force: Dynamics of the Earth*. Norton.

Westheimer, F. H. 1987. Why nature chose phosphates. *Science* 235: 1173–1178.

Wicken, J. 1987. *Evolution, Thermodynamics and Information: Extending the Darwinian Program*. Oxford University Press.

Wier, A. M., S. Luciano, M. F. Dolan, C. Bandi, J. MacAllister, and L. Margulis. 2010. Spirochete attachment ultrastructure: Implications for the origin and evolution of cilia. *Biological Bulletin* 218: 400–440.

Williamson, D. I. 1992. *Larvae and Evolution: Toward a New Zoology*. Chapman and Hall.

Williamson, D. I. 2001. Larval transfer and the origins of larvae. *Zoological Journal of the Linnean Society* 131: 111–122.

Williamson, D. I. 2003. *The Origins of Larvae*. Kluwer.

Williamson, D. I. 2009. Caterpillars evolved from onychophorans by hybridogenesis. *Proceedings of the National Academy of Sciences* 106: 19901–19905.

Williamson, D. I. 2012. The origins of larvae and the demise of Haeckelian zoology. In *Evolution from the Galapagos*, ed. G. Trueba. In press.

Williamson, D. I., and S. E. Vickers. 2007. The origins of larvae: Mismatches between the forms of adult animals and their larvae may reflect fused genomes, expressed in sequence in complex life histories. *American Scientist* 95: 509–517.

Wilson, E. O. 1991. Rain forest canopy: The high frontier. *National Geographic* 180: 78–107.

Winkler, W., A. Nahvi, A. Roth, J. A. Collins, and R. R. Breaker. 2004. Control of gene expression by a natural metabolite-responsive ribozyme. *Nature* 428: 281–286.

Woese, C. R., O. Kandler, and M. L. Wheelis. 1990. Towards a natural system of organisms, proposal for the domains Archaea, Bacteria, and Eucarya. *Proceedings of the National Academy of Sciences* 87: 4576–4579.

Zotin, A. I. 1972. Thermodynamic aspects of developmental biology. *Monographs in Developmental Biology* 5: 1–59.

Appendix A: Major Groups of Living Organisms

Source: Lynn Margulis and Michael J. Chapman, *Kingdoms and Domains: An Illustrated Guide to the Phyla of Life on Earth*, fourth edition (Elsevier, 2010), second printing

SUPERKINGDOM PROKARYA Origins not by symbiogenesis
  KINGDOM PROKARYOTAE (Bacteria, Monera, Prokarya)
    SUBKINGDOM (DOMAIN) ARCHAEA
      Division Mendosicutes (deficient-walled archaebacteria)
      Phylum B-1 Euryarchaeota methanogens and halophils
      Halobacter, Halococcus, Methanobacterium
      Phylum B-2 Crenarchaeota thermoacidophils
      Pyrobaculum, Pyrodictium, Sulfolobus, Thermoplasma
    SUBKINGDOM (DOMAIN) EUBACTERIA
      Division Gracilicutes (Gram-negative bacteria)
      Phylum B-3 Proteobacteria purple bacteria: phototrophs, heterotrophs
      Azotobacter, Escherichia, Nitrobacter
      Phylum B-4 Spirochaetae helical motile heterotrophs, periplasmic flagella
      Diplocalyx, Spirochaeta, Spirosymplokos
      Phylum B-5 Bacteroides-Saprospirae gliding fermenters, heterotrophs
      Bacteroides, Saprospira, Sporocytophaga
      Phylum B-6 Cyanobacteria oxygenic photoautotrophs
      Anabaena, Nostoc, Oscillatoria
      Phylum B-7 Chloroflexa gliding non-sulfur oxygen-tolerant photoautotrophs
      Chloroflexus, Heliothrix, Oscillochloris
      Phylum B-8 Chlorobia sulfur oxygen-intolerant photoautotrophs
      Chlorobium, Chlorochromatium, Chloronem

Division Tenericutes (wall-less eubacteria)
  Phylum B-9 Aphragmabacteria no cell walls
  Acholeplasma, Ehrlichia, Mycoplasma, Wohlbachia
Division Firmicutes (Gram-positive and protein-walled bacteria)
  Phylum B-10 Endospora low-G+C endospore-forming Gram-positives and relatives
  Bacillus, Clostridium, Peptococcus
  Phylum B-11 Pirellulae proteinaceous wall-formers and relatives
  Chlamydia, Gemmata, Pirellula
  Phylum B-12 Actinobacteria fungoid multicellular Gram-positives and relatives
  Actinomyces, Frankia, Streptomyces
  Phylum B-13 Deinococci radiation- or heat-resistant Gram-positives *Deinococcus, Thermus*
  Phylum B-14 Thermotogae thermophilic fermenters
  Aquifex, Fervidobacterium, Thermotoga

SUPERKINGDOM EUKARYA Origins by symbiogenesis
KINGDOM PROTOCTISTA
Four modes: phyla whose members
I. lack both undulipodia and meiotic sex
II. lack undulipodia but have meiotic sexual life cycles
III. have undulipodia but lack meiotic sexual life cycles
IV. have undulipodia and meiotic sexual life cycles
  SUBKINGDOM (Division) AMITOCHONDRIA
    Phylum Pr-1 Archaeprotista (III) motile with no mitochondria
    Mastigamoeba, Pelomyxa, Staurojoenina, Trichonympha
  SUBKINGDOM (Division) AMOEBAMORPHA
    Phylum Pr-2 Rhizopoda (I) vahlkampfid amoebae, cellular slime molds
    Acrasia, Arcella, Dictyostelium, Mayorella
    Phylum Pr-3 Granuloreticulosa (IV) reticulomyxids, foraminifera, chlorarachnids
    Allogromia, Globigerina, Rotaliella
    Phylum Pr-4 Xenophyophora (I) barite skeleton deep sea protists
    Galatheammina, Psammetta, Reticulammina
  SUBKINGDOM (Division) ALVEOLATA
    Phylum Pr-5 Dinomastigota (IV) dinoflagellates
    Gonyaulax, Gymnodinium, Peridinium
    Phylum Pr-6 Ciliophora (IV) ciliates

Gastrostyla, Paramecium, Tetrahymena
Phylum   Pr-7   Apicomplexa   (IV)   apicomplexan   animal symbiotrophs
Eimeria, Plasmodium, Toxoplasma
SUBKINGDOM (Division) HETEROKONTA (Stramenopiles)
Phylum Pr-8 Bicosoecida (III) small mastigotes, some form colonies
Acronema, Caféteria, Pseudobodo
Phylum Pr-9 Jakobida (III) bactivorous mastigotes. some loricate attached to sediment
Histiona, Jakoba, Reclinomonas
Phylum Pr-10 Proteromonadida (III) small mastigotes, intestinal in animals
Karotomorpha, Proteromonas
Phylum Pr-11 Kinetoplastida (III) kinetoplastids, most symbiotrophic mastigotes
Bodo, Crithidia, Trypanosoma
Phylum Pr-12 Euglenida (III) euglenids
Colacium, Euglena, Peranema
Phylum Pr-13 Hemimastigota (III) Gondwanaland mastigotes
Hemimastix, Spironema
Phylum Pr-14 Hyphochytriomycota (III) hyphochytrid water molds
Anisolpidium, Hyphochytrium, Rhizidiomyces
Phylum Pr-15 Chrysomonada (IV) chrysophytes, golden-yellow algae
Chrysosphaerella, Ochromonas, Synura
Phylum Pr-16 Xanthophyta (IV) yellow-green algae
Botrydium, Ophiocytium, Vaucheria
Phylum Pr-17 Phaeophyta (IV) brown algae
Ascophyllum, Fucus, Macrocystis
Phylum Pr-18 Bacillariophyta (IV) diatoms, silica tests
Diploneis, Melosira, Thalassiosira
Phylum Pr-19 Labyrinthulata (IV) slime nets and thraustochytrids
Althornia, Aplanochytrium, Labyrinthula
Phylum Pr-20 Plasmodiophora (IV) plasmodiophoran plant symbiotrophs
Ligniera, Plasmodiophora, Sorodiscus
Phylum Pr-21 Oomycota (IV) oomycete water molds (egg molds)

Achlya, Phytophthora, Saprolegnia
Phylum Pr-22 Amoebomastigota (III) amoebomastigotes
Naegleria, Paratetramitus, Willaertia
SUBKINGDOM (Division) ISOKONTA
Phylum Pr-23 Myxomycota (IV) plasmodial slime molds
Cercomonas, Echinostelium, Physarum. Stemonitis
Phylum Pr-24 Pseudociliata (III) polyundulipodiated animal
symbiotrophs
Opalina, Stephanopogon, Zelleriella
Phylum Pr-25 Haptomonada (III) haptophytes, coccolithophorids
Emiliania, Phaeocystis, Prymnesium
Phylum Pr-26 Cryptomonada (III) cryptomonads, cryptophytes
Chilomonas, Cryptomonas, Cyathomonas, Nephroselmis
Phylum Pr-27 Eustigmatophyta (III) eye-spot algae
Chlorobotrys, Eustigmatos, Vischeria
Phylum Pr-28 Chlorophyta (IV; green algae, plant ancestors)
Acetabularia, Chlamydomonas, Volvox
SUBKINGDOM (Division) AKONTA
Phylum Pr-29 Haplospora (II) haplosporan animal symbiotrophs
Haplosporidium, Minchinia, Urosporidium
Phylum Pr-30 Paramyxa (I) cell-inside-cell marine animal
symbiotrophs
Marteilia, Paramarteilia, Paramyxa
Phylum Pr-31 Actinopoda (II) ray animalcules; acantharia, helio-
zoa, radiolaria
Acanthocystis, Clathrulina, Sticholonche
Phylum Pr-32 Gamophyta (II) conjugating green algae
Micrasterias, Mougeotia, Zygnema
Phylum Pr-33 Rhodophyta (II) red algae
Amphiroa, Bangia, Gracilaria, Porphyra
SUBKINGDOM (Division) OPISTHOKONTA
Phylum Pr-34 Blastocladiomycota (IV) polyzoosporic water
molds
Allomyces, Blastocladiella
Phylum Pr-35 Chytridiomycota (IV; undulipodiated water molds,
fungal ancestors)
Monoblepharis, Neocallimastix, Polyphagus
Phylum Pr-36 Choanomastigota (IV; collared protists, animal
ancestors)
Dermocystidium, Nuclearia, Salpingoeca

KINGDOM ANIMALIA

SUBKINGDOM (Division) PLACOZOA (no nerves or antero-posterior asymmetry)

Phylum A-1 Placozoa free-living marine dorsal-ventral ciliated minimal animals

Trichoplax

Phylum A-2 Myxospora reduced marine fish symbiotrophs

Aurantiactinomyxon, Ceratomyxon, Sphaeromyxa

SUBKINGDOM (Division) PARAZOA (nerve nets)

Phylum A-3 Porifera sponges

Gelliodes, Leucosolenia, Spongia

Phylum A-4 Coelenterates (Cnidaria) sea anemones, hydroid-medusas

Craspedacusta, Hydra, Physalia

Phylum A-5 Ctenophora comb jellies

Beroë, Bolinopsis, Ocryopsis

SUBKINGDOM (Division) METAZOA (nervous and muscular systems; protostomatous adults, blastopore becomes mouth)

Phylum A-6 Gnathostomulida gnathostome worms

Austrognatharia, Gnathostomula, Problognathia

Phylum A-7 Platyhelminthes flat worms, flukes

Convoluta, Dugesia, Planaria

Phylum A-8 Rhombozoa tiny animal symbiotrophs

Conocyema, Dicyema, Microcyema

Phylum A-9 Orthonectida tiny animal symbiotrophs

Ciliocincta, Rhopalura, Stoecharthrum

Phylum A-10 Nemertina nemertine worms

Cerebratulus, Lineus, Prostoma

Phylum A-11 Nematoda nematode worms

Ascaris, Caenorhabditis, Rhabdias

Phylum A-12 Nematomorpha nematomorph worms

Chordodes, Gordius, Nectonema

Phylum A-13 Acanthocephala spiny-headed worms

Acanthogyrus, Leptorhynchoides, Macracanthorhynchus

Phylum A-14 Rotifera rotifers; wheel-animals

Brachionus, Euchlanis, Philodina

Phylum A-15 Kinorhyncha kinorhynch worms

Cateria, Echinoderes, Semnoderes

Phylum A-16 Priapulida priapulid worms

Halicryptus, Priapulus, Tubiluchus

Phylum A-17 Gastrotricha gastrotrich worms
Acanthodasys, Macrodasys, Urodasys
Phylum A-18 Loricifera loriciferans
Nanaloricus, Pliciloricus, Rugiloricus
Phylum A-19 Entoprocta entoprocts
Barentsia, Loxosoma, Pedicellina
Phylum A-20 Chelicerata spider, scorpions, ticks
Amblopygis, Ixodes, Scorpio
Phylum A-21 Mandibulata insects
Drosophila, Polistes, Pterotermes
Phylum A-22 Crustacea crayfish, lobsters, shrimp
Cambarus, Homarus, Linguatula
Phylum A-23 Annelida earthworms, polychaete worms
Hirudo, Lumbricus, Nephthys
Phylum A-24 Sipuncula sipunculid worms
Aspidosiphon, Phascolion, Themiste
Phylum A-25 Echiura echiurids
Bonellia, Metabonellia, Listriolobus
Phylum A-26 Mollusca clams, snails, squid
Busycon, Crepidula, Vema
Phylum A-27 Tardigrada water bears
Coronarctus, Echiniscus, Macrobiotus
Phylum A-28 Onychophora onychophorans
Cephalofovea, Peripatoides, Speleoperipatus
Phylum A-29 Bryozoa bryozoans, ectoprocts, moss animals
Bugula, Plumatella, Selenaria
Phylum A-30 Brachiopoda lamp shells
Crania, Lingula, Terebratulina
Phylum A-31 Phoronida phoronid worms
Phoronis, Phoronopsis
Phylum A-32 Chaetognatha chaetognath worms
Bathybelos, Krohnitta, Sagitta
Deuterostomatous adults (blastopore becomes anus)
Phylum A-33 Hemichordata acorn worms
Balanoglossus, Cephalodiscus, Ptychodera
Phylum A-34 Echinodermata starfish, sea urchins
Asterias, Solaster, Strongylocentrotus
Phylum A-35 Urochordata tunicates, ascidians
Ascidia, Halocynthia, Salpa
Phylum A-36 Cephalochordata lancelets
Amphioxus, Branchiostoma, Epigonichthys

Phylum A-37 Craniata vertebrates with skulls
Ambystoma, Cygnus, Pan
KINGDOM FUNGI
Phylum F-1 Microspora fish symbiotrophs
Glugea, Ichthyosporidium, Vairamorpha
Phylum F-2 Zygomycota mating molds
Pilobolus, Rhizopus, Trichomyces
Phylum F-3 Glomeromycota fish symbiotrophs
Acaulospora, Geosiphon, Gigaspora, Glomus
Phylum F-4 Ascomycota molds, morels, root symbiotrophs
Aureobasidium, Claviceps, Morchella, Saccharomyces
Phylum F-5 Basidiomycota mushrooms, puffballs, root symbiotrophs
Agaricus, Amanita, Phallus
Phylum F-6 Lichenes fungal + photosymbiotrophs
Cladonia, Letharia, Ramalina, Usnea
KINGDOM PLANTAE
Phylum Pl-1 Bryophyta true mosses
Bryum, Polytrichum, Takakia
Phylum Pl-2 Hepatophyta liverworts
Conocephalum, Marchantia, Porella
Phylum Pl-3 Anthocerophyta hornworts
Anthoceros, Folioceros, Megaceros
Phylum Pl-4 Lycophyta club mosses
Lycopodium, Phylloglossum, Selaginella
Phylum Pl-5 Psilophyta psilophytes
Psilotum, Tmesipteris
Phylum Pl-6 Sphenophyta horsetails
Equisetum
Phylum Pl-7 Filicinophyta ferns
Azolla, Osmunda, Polypodium
Phylum Pl-8 Cycadophyta cycads; sago palms
Ceratozamia, Cycas, Macrozamia
Phylum Pl-9 Ginkgophyta maidenhair tree
Ginkgo
Phylum Pl-10 Coniferophyta conifers
Abies, Pinus, Tsuga
Phylum Pl-11 Gnetophyta gnetophytes
Ephedra, Gnetum, Welwitschia
Phylum Pl-12 Anthophyta flowering plants
Aster, Liriodendron, Oenothera

Appendix B: The International Geological Time Scale (Time-Rock Divisions)

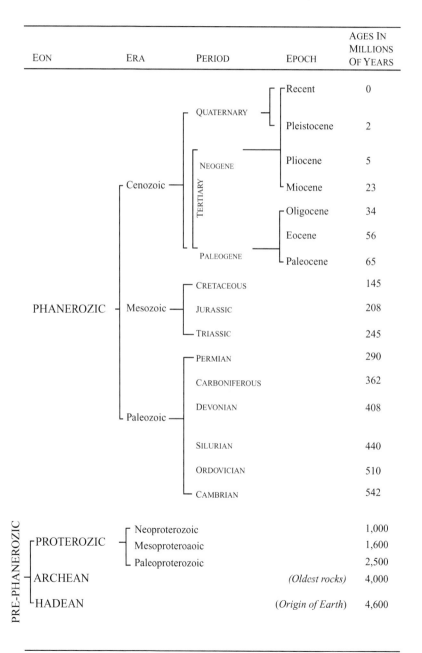

EON	ERA	PERIOD	EPOCH	AGES IN MILLIONS OF YEARS
PHANEROZIC	Cenozoic	QUATERNARY	Recent	0
			Pleistocene	2
		NEOGENE (TERTIARY)	Pliocene	5
			Miocene	23
		PALEOGENE (TERTIARY)	Oligocene	34
			Eocene	56
			Paleocene	65
	Mesozoic	CRETACEOUS		145
		JURASSIC		208
		TRIASSIC		245
	Paleozoic	PERMIAN		290
		CARBONIFEROUS		362
		DEVONIAN		408
		SILURIAN		440
		ORDOVICIAN		510
		CAMBRIAN		542
PRE-PHANEROZIC	PROTEROZIC	Neoproterozoic		1,000
		Mesoproteroaoic		1,600
		Paleoproterozoic		2,500
	ARCHEAN		*(Oldest rocks)*	4,000
	HADEAN		*(Origin of Earth)*	4,600

(Not to scale)

Glossary

abiotic   Nonliving

acetyl coenzyme A   Ubiquitous metabolic intermediate in lipid biosynthesis derived from pathways such as glycolysis, fatty-acid oxidation, and amino acid degradation.

aconitase   Citrate-induced enzyme repressed by glutamate.

adaptive radiation   Evolution of different varieties or more inclusive taxa (e.g., species, genera from common ancestors.

adenylyl cyclase   Enzyme that catalyzes the formation of cyclic AMP.

aer   Transducing protein involved with aerotaxis.

alarmones   Stress-induced ribonucleotide derivatives; hypothesized to have been synthesized in RNA/protein cells before the evolution of DNA genomes.

allelo chemicals   Ecological term that refers to chemicals that influence behavior or development when released into the environment.

amination   Chemical process by which an amine group ($-NH_2$) is introduced into an organic molecule.

amino acid   Any one of more than twenty nitrogenous monomers that may align in peptide bonds to form proteins.

amygdala   Brain center of the limbic system involved in emotional conditioning and communicative signals relating to fear and threat.

anaerobic respiration   Metabolic mode requiring electron transport chains in which inorganic compounds, but not oxygen, acts as terminal electron acceptor and therefore is reduced (e.g., nitrate, sulfate, elemental sulfur).

anaphase   Stage of mitosis (eukaryotic cells) in which half-chromosomes (chromatids) segregate at their centromere/kinetochores and move on spindle microtubules toward opposite poles.

anhydride   Chemical compound derived from another by removal of water.

apoptosis   Genetically programmed or scheduled cell death.

ApppA   Bacterial alarmone; abbreviation of diadenosine triphosphate.

archaea   Archaebacteria. Prokaryotic microorganisms with distinctive 16S ribosomal RNA sequences, cell walls, and membranes. See appendix.

archaeprotists   Phylum of kingdom Protoctista comprised primarily of anaerobic amitochondriate or microaerophilic protists, including amoebas and mastigotes. See appendix.

Archean eon   Earlier of two time-rock divisions of the Precambrian that spans a period from 3,900 million to 2,500 million years ago.

ascidian   Sedentary marine animal, phylum Urochordata of class Ascidiacea with a sack-like body and siphons through which water enters and exits.

associative cortex   Isocortical regions of craniate brains involved in the processing and integration of complex, multimodal information.

ATP   Adenosine triphosphate; a ring compound ubiquitous in metabolic pathways of life on Earth used to store and release energy by forming or breaking its phosphate bonds.

autoinducer   Biologically active compound, a pheromone produced by and responded to by organisms of the same population.

autoinduction   Cell-to-cell signaling system. See also quorum sensing.

autopoiesis   Process of identity and self-maintenance of life, a behavior characteristic of the living. Sum of metabolic and reproductive activities of organisms responsible for self-bounding, self-maintenance, and self-propagation.

axostyles   Axial motility organelle of archaeprotist mastigotes, metamonads, and parabasalians composed of a patterned array of microtubules and their cross-bridges. Axostyles extend from the anterior toward the posterior end of the cell.

bacteriorhodopsin   Isoprenoid-derived pigment that absorbs light and generates a proton gradient across the membrane of salt-requiring halobacteria in photosynthesis by photophosphorylation of ATP without chlorophyll.

basal ganglia   Brain structures; bundled cell bodies of neurons involved in voluntary, stereotyped behavior motor learning, and motivation.

base pair   Monomer components of nucleic acids, the hydrogen-bonded complementary deoxyribonucleotide molecules (DNA) and (RNA); examples include AT (2H bonds) adenine-thymine or GC (3H bonds) guanine-cytosine ribonucleotides.

biofilm   Three-dimensional flat benthic communities of microbes; primarily mixed populations of bacteria that produce a self-developed matrix and adhere to a living or inert surface. Complex community interactions occur inside; the structure is a product of lipids, proteins, and other polymeric substances.

bioluminescence   Emission of light by microorganisms such as the bacteria (*Vibrio fischeri*) or protists marine dinomastigotes (*Noctiluca*) when population densities reach a threshold that stimulates a concentration of auto-inducer.

biosphere   "Sphere" near Earth's surface in which life exists, from a depth of 12 kilometers in the oceans to an altitude of 8 km in the atmosphere.

biospheric light   Portion of the electromagnetic spectrum from 200 to 1,100 nanometers, to which organisms respond (including polarized light). Humans react to a more limited (narrower) range: 400–700 nm.

brachiopods   Marine animals, phylum Mandibulata; with bivalve shell and a pair of tentacle-like arms which serve to capture food.

Cambrian period   Period on Paleozoic era from 542 million to 488 million years ago, noted for sudden appearance of animals with hard parts.

capsid   Protein coat of virus.

carotenoids   Isoprenoid pigments in plastids of virtually all phototrophic organisms and many heterotrophs.

cathemerality   Behavioral activity of animals in which activity is distributed almost evenly throughout the dark and the light portion of the 24-hour period, distinguished from diurnal and nocturnal.

centriole   Intracellular proteinaceous barrel-shaped organelle consisting of longitudinally aligned microtubules where nine groups of triplet peripheral microtubules surround a central cavity. Pattern of microtubules is described as [9(3) + 0]. A kinetosome is the same structure from which an axoneme extends distally.

centromere   Specialized region of chromosomal DNA (the nucleic acid portion) that joins two half-chromosomes (chromatids) in mitotic cells. The proteinaceous portion at the same location that joins two half-chromosomes (chromatids) is called the kinetochore.

centrosome-associated RNA or cnRNA sequences   RNA molecules that bind basic proteins and enhance nucleation of microtubules.

centrosomes   Central regions of animal cells (e.g., eggs), external but adjacent to the nucleus, that reproduce and generate dynamic spindle microtubules synthesized and broken down during mitotic cell division.

cerebellum   Portion of the mammalian brain, situated above the medulla oblongata and beneath the cerebrum, that is important to motor control.

cetaceans   Order of aquatic mammals that includes whales, dolphins, porpoises, and narwhals; characterized by fusiform hydrodynamic bodies, fin-like forelimbs, and the absence of hind limbs.

cGMP   Cyclic guanosine monophosphate; compound regulates metabolic processes and may be an antagonist to cyclic AMP.

chemoautotrophy, chemolithoautotrophy   "Self-feeding" mode of nutrition limited to prokaryotes in which the energy source is an oxidation of reduced inorganic compounds; e.g., H_2S or sulfur to sulfate; ammonia to N_2 or NO_3; hydrogen (H_2) to (H_2O), inorganic compounds. Such reduced inorganic compounds provide sources of electrons in fixation of carbon dioxide (CO_2) to organics.

chemosensitivity   Sensitivity to changes in the chemical composition of the surrounding fluid (gas, liquid).

chemostat   Container that permits continuous growth of a pure culture bacterial population. A growth-rate limiting resource is added at the same rate that depleted medium and cells are washed out. The bacterial population consumes the resource until density equilibrium is reached and the population's growth rate equals the resource flow rate through the apparatus.

chemotaxis Movement (swimming or gliding) of a microbe toward or away from a chemical stimulus (chemotactic agent) such as a concentration gradient of food source (positive chemotaxis) or toxin (negative chemotaxis).

chimera Organism generated by merger of two or more different genetic types formed by symbiogenesis, hybridogenesis, or other mode of fusion.

chirality Molecular "handedness." Molecules lack an internal plane of symmetry, and differently rotate the plane of polarized light that cannot be superimposed on their mirror image (e.g., most amino acids in life are levorotatory and most sugars are dextrorotary).

chromatin Eukaryotic DNA/histone complex; DNA-nucleosomes that forms the condensed chromosomes during mitotic cell division.

chromatophore Pigment-containing structure or organelle that forms the colored portion of an organism or a cell.

chromonema Coiled central filament of a chromatid upon which the chromomeres rest.

chromosome-chromolinkers Bridges between chromosomes that connect all the chromosomes of a haploid set into a single continuous DNA-linked structure.

ciliate Phylum of protoctists with an estimated 10,000+ species in aquatic environments. Most are single-celled and have dimorphic nuclei.

circadian rhythms Diurnal-nocturnal physiological phenomena (e.g., cell division photosynthetic rate, bioluminescence intensity, cyclical, and enzyme production) that occurs with a periodicity of approximately 24 hours.

cladogram Diagrammatic representation of a "tree" topology of assumed monophyletic evolutionary relationships of taxa.

coenzyme A (CoA) Sulfur compound synthesized from vitamin B required for polymerization of acetates to form fatty acids in lipid metabolism.

colligative Density-dependent properties of solutions as opposed to properties contingent upon characteristics of the molecules in solution. Colligative properties include: lowering of vapor pressure; elevation of boiling point; depression of freezing point and osmotic pressure.

Cretaceous Final period of the Mesozoic Era, from 145 million to 65 million years ago.

cyclic AMP Ubiquitous nucleotide: cyclic adenosine monophosphate (cAMP) synthesized from adenosine triphosphate (ATP); used for intracellular signal transduction.

cysteine Sulfur-containing amino acid in most proteins; oxidizes to cystine upon exposure to oxygen.

cytoplasm Fluid portion of cells that contains enzymes and metabolites in solution.

cytoskeleton Microfilaments, microtubules, and their associated proteins form an asymmetric, dynamic scaffolding associated with cell motility; limited to eukaryotic cells.

decarboxylation   Removal of CO_2 (one or more carboxyl groups, -COOH) from a molecule.

determinism   View that every event, from behavior to cognition, is determined by a continuous chain of causal events.

diatoms   Members of the phylum Bacillariophyta in the Kingdom Protoctista; unicellular and colonial aquatic protoctists covered by two-part silica (SiO_2) bivalve shells called tests.

diencephalon   Posterior part of the forebrain that contains the thalamus and hypothalamus; involved in regulation of visceral function.

differentiation of self (DoS)   Murray Bowen's evaluation of human emotional behavior, a measurement of a person's capacity to appropriately perceive and respond directly to the immediate physical and social environment (high value) compared to "reactivity" where a person's perceptions of and responses to physical and social stimuli are unconsciously based on his/her past experience (low values).

dimer   Chemical compound that consists of two structurally homologous subunits (monomers) bonded together.

dinomastigotes (dinoflagellates)   Members of the phylum Dinomastigota; biundulipodiated; most single-celled and testate planktic marine protoctists; many photosynthetic.

divalent metal ion catalysis   Biochemical process that speeds up reactions by use of proteins with metal cations (e.g., Mn^{2+}, Mg^{2+}, Fe^{2+}) that activate and bind H_2O. They liberate one hydrogen atom. Such reactions may produce a proton gradient force used to catalyze other reactions.

echinoderm   Member of the Phylum Echinodermata and the Kingdom Animalia; a marine animal (e.g., a sea urchins or a starfish) with an internal calcareous skeleton.

eclogite   Course-grained igneous or metamorphic rock; similar to basalt composed primarily of garnet and clinopyroxene.

electron donor   Reducing agent; a chemical compound in metabolism that simultaneously becomes oxidized when it transfers electrons that reduce another compound.

electron receptor   Oxidizing agent that simultaneously becomes reduced when it accepts electrons from any other compound.

embryophyte   Member of the Kingdom Plantae. (See appendix.)

endolithic and epilithic lichens   Lichens that live in (endo-) or on (epi-) rocks.

energy core   Universal metabolism, i.e., multi-enzyme catalyzed metabolism that yields energy present in all life on Earth and inferred to have been present in the last common ancestor.

enol   Organic chemical with a hydroxyl group (-OH) bonded to two double-bonded carbon atoms.

equilibrium chemistry   Dynamic chemical balance in which the rate of product formation from the reactants equals the rate of reactant formation from the products.

Eubacteria   The largest by far of the subkingdoms (or domains) of the Kingdom Prokaryotae.

Eukarya   Superkingdom: formal name of taxon that includes all nucleated (eukaryotic organisms).

exocytosis   Cell secretion in eukaryotes involves intracellular motility and membrane traverse.

exopolymeric substances   Biopolymers, polysaccharids, and/or proteins secreted into the environment.

facultative   Optional; e.g., facultative autotrophs grow either autotrophically or heterotrophically.

fatty acid synthesis   Production of lipid or long-chain aliphatic carboxylic acid of fats and oils inside cells.

ferredoxins   Nearly universally distributed iron- and sulfur-containing proteins that function in photosynthetic and respiratory electron transport.

flagella   Bacterial flagellum, prokaryotic extracellular structure made of polymers of flagellin proteins that move by rotation at the base of a relatively rigid rod, driven by a rotary motor embedded in the cell membrane; intrinsically nonmotile, and sometimes sheathed. Undulipodia, mistakenly called flagella, differ in every aspect from prokaryotic flagella; they are intrinsically motile intracellular structures.

flavins   Class of organic compounds that includes yellow-pigmented compounds, such as riboflavin.

flavonoid   Class of plant pigments with a chemical structure similar to flavones.

foraminifera   Members of the largest, most diverse class in Phylum Reticulomyxa of Kingdom Protoctista; reticulating pseudopods; marine single cells; often huge with tests.

fovea   Retinoid depression in some vertebrates; the point where the vision is most acute.

formate   Organic compound: any salt or ester of formic acid (HCOOH).

fossorial   Adjective pertaining to digging animals that live in subterranean habitats.

functional group   Ecology: Communities of organisms grouped by morphological, physiological, behavioral, biochemical, or environmental responses, or on trophic criteria. Chemistry: Atomic groups in molecules confer chemical properties on those molecules. Functional groups react similarly regardless of the molecules to which they are attached.

Gaia   Modulation by life of Earth's temperature, acidity, and reactive atmospheric gases and solutions through dynamic interaction of organism and the planet's surface.

genome   Total genetic makeup of an organism, including all its DNA.

genophore   Gene-bearing structure of prokaryotes and organelles of bacterial origin; should replace the outdated term "bacterial chromosome." Nucleoids are genophores visible by electron microscopy. Genophores are nucleoid equivalents inferred from genetic investigations.

geomicrobiology   Scientific discipline that studies microbial processes in geological and geochemical phenomena.

gliding motility   Movement of cell or organism always in contact with a solid surface in the absence of external appendages.

gnathostome   Animals in phylum Craniata, subphylum Vertebrata that possess jaws. (See appendix.)

gradient   Difference across a distance in forces or conditions; e.g., temperature, pressure, or chemical oxidation state.

gravisensors   Intracellular sensory organelles or organs that determine an organism's response to gravity.

Hadean Eon   The eon from Earth's formation (4,600 million years ago) to its oldest dated rocks (>3,900 mya).

haplorhini   Mammalian taxon in the Order of Primates that includes prosimian tarsiers and true simians.

heterocysts   Nonphotosynthetic cells of filamentous cyanobacteria that synthesize nitrogenase that enables "nitrogen fixation." Inert atmospheric nitrogen (N_2) is converted into acceptable N-containing food molecules.

hippocampus   Structure of mammalian brains involved in motivation, emotion, and memory.

histocompatiblity genes   Set of genes that encode proteins important to antigen formation in the immune response.

histone   Lysine- and arginine-rich protein that complexes with DNA in eukaryotes in formation of the nucleosomes of chromatin. Histone quantity per cell is proportional to ploid (haploid cells have half the histone content of diploid cells) is similar in plants, animals, and fungi, but greatly variable in protoctists.

holobionts   Evolutionarily stable symbiotic associations of bionts (symbionts). All eukaryotes are holobionts.

hominids   Animals in Class Mammalia, order Primates, family Hominidae.

hydrogenases   Enzymes that catalyze the reversible oxidation of molecular hydrogen (H_2) and produce H_2 gas and protons.

Hymenoptera   Order in Phylum Mandibulata, Class Hexapoda (Insecta) of Kingdom Animalia in which sawflies, wasps, bees, and ants are placed.

immunoglobin   Any glycoproteins of blood serum; an antibody produced in response to unfamiliar antigens. These proteins protect animals by removal or immobilization of antigens.

integrase   Enzyme produced by retroviruses that enables retroviral DNA to be integrated into animal nuclear DNA.

integument Outer protective covering of organism, e.g., skin, shell, rind, cuticle, or seed coat.

interneurons Type of animal nerve cell; a neuron of the central nervous system that form a pathway.

intramolecular cyclization Reaction in organic chemistry that converts a linear molecule into a ring-shaped molecule.

ion Atom or group of atoms with electric charge, positive or negative, due to gain or loss of one or more electrons.

ion channels Pore-forming proteins in membranes of all cells that control trans-membrane voltage gradients; they allow or prevent flow of ions through membranes.

isocortex Mammalian brain structures, the six-layered, outer portion of the cerebral hemispheres; involved in sensory perception reasoning, conscious thought, and language.

isoprenoids Organic compounds with a structure based on repetitions of an ubiquitous five-carbon compound precursor (isopentenyl pyrophosphate).

karyomastigont Organellar system that evolved into the mitotic spindle comprising nucleus, proteinaceous nuclear connector, and undulipodium (as kineto-some/centriole + axoneme); an ancestral feature of eukaryotic cells.

kinetosomes Intracellular proteinaceous microtubular organelle characteristic of eukaryotic cells; 9(3) + 0 in transverse section; homologous to the centriole from which an axoneme extends.

kinocilium Specialized mechanoreceptive undulipodium on the apex of cells in the sensory epithelium of the vertebrate inner ear.

kinetochore Proteinaceous specialized region of the chromosome that joins the two half-chromosomes (chromatids) bound to centromeric DNA.

latent information Embedded data in the environment that, once processed cognitively, initiates some form of change in the sensing organism.

L(U)CA Last (Universal) Common Ancestor—the most recent organism from which theoretically all extant organisms on Earth descended.

Lepidoptera Order in Phylum Mandibulata, Class Hexapoda (Insecta) of Kingdom Animalia in which moths and butterflies are placed.

limbic system Brain structural system comprising many areas that effect control of the autonomic nervous system and emotional states.

lipid bilayer Structural molecular array phospholipid molecules that form cell membranes.

lipids Organic compounds (e.g., fats, waxes, and steroids) that are soluble in organic solvents (alcohols, acetone) but not in water.

lux genes Genes that code for luciferase enzymes in bacteria that produce "cold light" (bioluminescence).

lytic enzymes Proteins that enzymatically break down organic materials (e.g., that destroy cell membranes by cutting or "lysing").

magnetosomes Prokaryotic intracellular organelles of bacteria that contain single domain magnetite crystals by which magnetotactic bacteria orient themselves in geomagnetic fields.

mechanosensitivity Sensitivity to mechanical stimuli such as pressure or touch.

Mecoptera Order in Phylum Mandibulata, Class Hexapoda (Insecta) of Kingdom Animalia in which "scorpionflies" are placed.

medulla oblongata Mammalian brain structure, that with the pons, constitutes the hindbrain, which controls heart rate, constriction and dilation of blood vessels, respiration, and digestion.

meiosis One or two successive divisions of a diploid nucleus in which the number of chromosomes is reduced by half, leading to haploid offspring cells.

membrane enzymes Catalytic proteins embedded and usually anchored within a cell membrane.

meme Term, coined by the biologist Richard Dawkins, referring to a unit of cultural transmission analogous to genes claimed to self-propagate and "evolve" through a population.

Mendelian genetics Rules of transmission of hereditable characteristics passed from parents to their offspring (e.g., in diploid eukaryotic organisms such as animals and plants). Mendelian genetics does not apply to prokaryotic transmission genetics.

mercapto groups In chemistry, univalent functional group –SH or mercapto radical. Thiols are organic compounds that contain them.

mesencephalon Brain structure of vertebrates composed of the dorsal tectum and ventral tegmentum.

mesosomes Invaginations of the plasma membranes of prokaryotes (Gram-positive and Gram-negative bacteria) that form vesicles thought to be involved in cell wall formation. They may be artifacts of preparation of bacteria for electron microscopy.

metabolites Organic compounds produced by or active in metabolic processes of organisms.

metagenomics Study of genetic material (DNA) extracted directly from environmental samples.

metaphase Stage in division (mitosis) in eukaryotic cells during which visible chromosomes align at the equator prior to separation into two offspring cells.

metaphysical materialism Concept that everything is either body or matter; all that exists is either physical (material) or wholly dependent upon the physical.

methanogenesis Production of methane (CH_4) by the metabolic activity of methanogenic prokaryotes, members of the subkingdom (domain) Archaebacteria.

methyl-accepting chemotaxis proteins (MSPs) Bacterial sensory molecules that methylate (bind to –CH_3 groups) in response to an increased concentration of a particular substance, typically a toxin or energy source.

microbial IQ Ability of a prokaryote to sense and respond to environmental change; defined in part by the number of sensory systems found in its genome.

microbe Any organism not visible to the unaided eye.

microbialites Organic sedimentary deposits made by benthic (bottom dwelling or lowest region of a body of water) microbial communities.

microbiota Sum of the microorganisms in a given habitat; replaces "microflora" and "microfauna," which imply they are plants and animals, respectively.

Mid-Proterozoic (Mesoproterozoic) Eon Time-rock division that ranges from 1,600 million to 1,000 million years ago, comprising the Calymmian, Ectasian, and Stenian periods. This eon was related to the formation of the first super continent mountain building events, abundance of stromatolites, and major changes in oceanic and atmospheric chemistry.

mitochondria Intracellular organelles in nearly all eukaryotic cells that synthesize ATP by use of O_2 gas as terminal electron acceptor. Probably began as oxygen-respiring purple bacteria that were acquired symbiogenetically and resisted digestion after incorporation into amitochondriate mastigotes.

mitosis Eukaryotic nuclear division that produces two genetically identical offspring cells with alternation in chromosome number.

mononucleotide Single nucleotide.

motif Pattern of sound constituting an animal call.

multicellularity Individual organisms comprising more than a single cell. Multicellularity evolved independently in all five kingdoms.

myxobacteria "Slime bacteria," eubacteria in Phylum Proteobacteria of Kingdom Prokaryotae. Formal name of the bacterial kingdom, the higher most inclusive taxon that includes two subkingdoms (domains: Archaea = Archaebacteria and Eubacteria).

myxospores Microscopic symbiotrophic animals that belong to the myxozoan phylum recently transferred to the animal kingdom. They have a complex life history as symbiotrophs in invertebrates and ectothermic vertebrates.

NAD Nicotinamide adenine dinucleotide, a nearly ubiquitous coenzyme electron acceptor in respiration; derived from nicotinic acid from the B vitamin.

NADP Nicotinamide adenine dinucleotide phosphate, an electron acceptor and donor in electron transport; reduced in light reactions of photosynthesis.

naive realism Common-sense realism; perception corresponds accurately with the external world.

necrotrophy Nutritional mode in which a symbiotroph damages or kills the live organism that is the source of its nutrient.

neo-Darwinism "Modern" or "new" synthesis of twentieth-century concepts that unites stability (genetic reassortment without evolutionary change) intrinsic to Mendelian genetics with Darwinian evolution (change through time) in which random genetic mutation is taken to be the major source of heritable novelty upon which natural selection acts.

nepheloid layer Water that contains subaqueous suspended sediment and organic matter immediately above sea or lake bottom strike.

nerve-cell morphogenesis Growth and maturation in animal bodies of undifferentiated epithelial cells into neurons with axon and dendrites.

neural plasticity Ability of brain or other nervous system tissue to differentiate new functions in response to new environmental conditions.

nicotinamide Amide of niacin (vitamin B_3), precursor to co-enzyme NAD or NADP; a major component of biochemical pathways of cellular respiration.

nitrogenase Enzyme system in use by nitrogen-fixing prokaryotes that catalyze conversion of nitrogen gas to ammonia.

nitrogen fixation Crucial metabolic pathway limited to certain prokaryotes essential to the biosphere; the reduction of inert dinitrogen atmospheric gas (N_2) to usable ammonia (NH_3) by enzyme system called nitrogenase.

nodulation Formation of spherical bodies or nodules (e.g., in plant roots that fix nitrogen) because of symbiotic association with N_2-fixing bacteria.

non-equilibrium thermodynamic systems Systems that are neither isolated or closed off from their environments with respect to matter and energy flow and therefore do not reach thermodynamics equilibrium and correlated lack of change. Natural systems in a state of non-equilibrium that can do work as they are subject to continuous matter and energy exchange in which heat flows into the cool.

nucleated cells Eukaryotes with membrane-bounded nuclei—e.g., plant, animal, fungal, and protoctist cells.

nucleoid DNA-containing structure in prokaryotic cells visible by thin-section electron microscopy in favorable preparations. Nucleoids are not bounded by pore studded membrane.

nucleolinus Organelle of eukaryotes; dynamic RNA-rich body in the nucleolus that acts as microtubule organizing center for centrosomes and mitotic spindle in large clam eggs.

nucleotides Monomers of the nucleic acids DNA or RNA.

oligomer condensation Chemical reaction unit that binds two activated mono-nucleotides to form nucleic acid oligomers.

oncolites Small, variously shaped, concentrically laminated calcareous rocks formed by successively layered masses of cyanobacteria.

onychophorans Members of Phylum Onychophora in the Kingdom Animalia; bilaterally symmetrical wormlike carnivores common in most soil and leaf litter of tropical and temperate forests; peripatus or velvet worms.

ooids Spherical or ellipsoid concretions of calcium carbonate; sand-sized sedimentary structures produced by microbial communities.

oolites Rocks that form ovoid or spherical deposits have a concentric or radial structure; most made of calcium carbonate, but some of calcium phosphate, silica, iron silicate, iron oxide or iron carbonate, and siderite.

opisthobranchs   Members of the Phylum Mollusca of the Kingdom Animalia and Class Gastropoda, including small sea snails, bubble snails, sea slugs, and many other nudibranch marine mollusks with asymmetrically coiled shells. Some photosynthetic nudibranchs evolved as the internal cytoplasm of green algae and retained their chloroplasts. Feeding heterotrophs that ingested the internal cytoplasm of green algae and retained their chloroplasts.

osmotic pressure   Water pressure exerted by a solution across a semi-permeable membrane by a solvent.

oxygenase   Enzyme that catalyzes biochemical reactions involving molecular oxygen (O_2).

oxygenic photosynthesis   Mode of nutrition that uses light energy to split H_2O and produces a proton gradient by liberation of hydrogen ions and release of gaseous oxygen (O_2) as waste. Only cyanobacteria, algae, and plants are capable of oxygenic photosynthesis.

pericentriolar material   Visible in electron micrographs, this proteinaceous fuzz in the cytoplasm of nucleated cells (e.g., eggs) is associated with centrioles or centrosomes.

phagocytosis   Mode of heterotrophic nutrition limited to eukaryotes; ingestion of microbes or solid food by amoeboid cells; also immunological defense in animals where pseudopods engulf and destroy foreign particles.

Phanerozoic eon   Time-rock division that ranges from 542 million years ago to the present. Because of the abundant evidence of Phanerozoic macroscopic life, this period of geological time marks the classical fossil record.

phase relationship   Description of time relations. A temporal connection or synchronization between two or more rhythms.

pheromone   A chemical substance that when released into the environment correlates with changes in interspecific behavior or development. If the chemical is produced in a given species by one gender and specifically affects the other gender, it is called a sex pheromone.

phosphate carrier   Any phosphate-containing organic compound in metabolism that can be deployed such that the phosphate group detaches and becomes available for use in another pathway or organism.

phosphoenolpyruvate   Small, abundant organic intermediary compound in energy core metabolism and glycolysis; perhaps ubiquitous in life.

phosphoenolpyruvate transfer   Step in glycolytic and energy core metabolism in which pyruvate kinase catalyzes transfer of a high-energy phosphate group from phosphenolpyruvate to ADP to produce ATP and pyruvate.

phosphoglycerates   Salts or esters of phosphoglyceric acid used in the energy-growth core metabolism.

phospholipids   Primary constituents of the cell membranes of living organisms; lipids; glycerol molecules bound to fatty acid chains that have phosphate groups attached.

phosphoric acid H_3PO_4, an inorganic acid.

phosphorylation   Metabolic reaction that adds a phosphate group (PO_4) to organic molecules. Removal of phosphate groups is dephosphorylation.

photoautotrophy   Mode of nutrition in which light acts as the energy source by inducing conformational changes in photosensitive molecules (pigments) that then catalyze metabolic reactions that build up cell material from CO_2. Obligate photoautotrophs use light energy to synthesize their cellular components from carbon dioxide without any organic carbon, not even vitamins.

photobiology   Science of the interaction of light and life, including photosynthesis, bioluminescence, and vision.

photosensitive   Responding to exposure to photons, particularly visible light.

photosynthesis   Any of several light-sensitive metabolic modes in which production of cell material is enhanced by light (e.g., anoxygenic photosynthesis in *Chlorobium* or halobacter; photoautotrophy in food plants).

photovoltaic effect   Property of certain chemical elements when exposed to electromagnetic radiation (i.e., in the visible light region of the spectrum); a transfer and subsequent accumulation of electrons occurs in the material that produces a voltage.

phycocyanin, phycoerythrin   Light-sensitive proteins bind to a blue-green colored pigment (phycobiliprotein) of cyanobacteria, and plastids of certain algae acquired from cyanobacteria (e.g., rhodophytes and cryptomonads) that captures light energy and transfer it to chlorophyll pigments in photosynthesis.

phylotype   Evolutionary label on a microbe based primarily on sequence comparisons of 16S rRNAs.

phytochrome   Photosensitive pigment of plants that detects light and may influence stem, seed or leaf growth, alter photoperiodism, and change orientation of chloroplasts within a plant cell to maximize light absorption.

pisolites   Spherical pea-sized or ovoid rocks composed primarily of calcium carbonate and larger than 2 millimeters in diameter.

plague culling   Aggressive symbiosis in which a necrotrophic bacterium maims or kills a genetically susceptible subset of an animal population. Selection for survival of a genetically less-susceptible population that then co-evolves with the bacterium, process tends to maintain symbiotic associations.

Planctosphaera   Monospecific genus of spherical planktic animals with one extant species (*Planctosphaera pelagica*). Classified in the Phylum Hemichordata and Kingdom Animalia with acorn worms and pterobranchs because of its resemblance to the larvae of these proboscis-bearing bilaterally symmetrical soft bodied marine worms.

Pogonophorans   Members of the animal phylum Pogonophora that are sessile, marine benthic chitinous-tube-forming worms (beard worms).

polydnavirus   Virus fully incorporated into the genome of separate species of wasps. Upon laying of eggs by a wasp into a live victim, the virus suppresses the victim's immune system.

polymerase Enzyme that catalyzes the formation of polymers from monomers; e.g., DNA polymerase synthesizes DNA from deoxynucleosides.

polypeptide Amino acid polymer joined via carbon-nitrogen chemical links called peptide bonds. All proteins are long chain polypeptides.

pons Brain structure that acts as to relay nerve impulses between the cerebellum and the cerebrum and functions in control of breathing.

ppGpp and pppGpp Guanosine pentaphosphate and tetraphosphate, respectively, small phosphate-rich organic compounds, alarmones, that inhibit RNA synthesis during amino acid shortage.

prefrontal cortex (PFC) Brain structure in the prefrontal or anterior part of the frontal lobe of the cerebral cortex of mammals associated with abstract thought, social behaviors, planning, and decision making.

prokaryote Any member of the Kingdom Prokaryotae in either subkingdom (domain) Eubacteria or subkingdom (domain) Archaea (Archaebacteria) composed of one or many small ribosomal, genophore-bearing prokaryotic cells.

prometaphase Stage of mitotic cell division in animals in which the nuclear membrane disintegrates, the centrioles move to the poles of the cell and the chromosomes continue to condense.

propagule Genome-containing non-growing cell or multicellular structure capable of survival, subsequent dissemination, growth and maturation into an adult, reproductive organism, (e.g., bacillus spores, actinospores, zoospores, basidiospores, seeds, tardigrade tuns, bryozoan, statoblasts, etc.).

proteins Organic polymers made up of long chains of peptide-bonded amino acids (e.g., serine, glycine, alanine, aspartic acid).

proteobacteria Kingdom Prokaryotae, subkingdom (domain) Eubacteria to which an immense diversity of Gram-negative unicellular and multicellular bacteria are assigned.

proteolysis Degradation of proteins into peptides and amino acids through the hydrolytic breakage of their peptide bonds by various enzymes called proteases.

Proterozoic Eon Time-rock division from 2,500 million years ago to 542 mya, notable for absence of animal and plant fossils.

protocells Self-organized, endogenously ordered, spherical accumulations of organic materials proposed as a step in the origin of life.

protoctists (informal English name of protoctista) One of the five kingdoms (highest or most inclusive taxa of life. Domains: Archaebacteria, Eubacteria, Eukarya are equivalent to Subkingdoms in biological classification). The Protoctista Kingdom includes smaller single-or few-celled eukaryotes, the protists and their syncitial or multicellular descendants. Eukaryotic organisms exclusive of animals, fungi and plants are protoctists. An estimated 250,000 extant protoctists species are classified into approximately 50 phyla, e.g., Ciliophora (ciliates), Chlorophyta (green algae) and Rhodophyta (red seaweeds).

proton-motive force Potential energy stored as an electrochemical gradient that is generated by the pumping of ions across cell membranes (e.g., in chemiosmosis).

prototaxis Innate tendency of one type of organism or cell to respond in a predictable way to another type of organism or cell.

punctuated evolution Model for evolution in which short periods of rapid evolutionary change are interspersed with long periods of evolutionary stasis.

pure culture Growth medium that contains a single kind of microorganisms.

pyruvate Intermediary compound (H_3–CO–COOH) in core metabolic pathway that is converted to acetyl coenzyme A, used in lipid biosynthesis, citric acid cycle for amino acid biosynthesis, and produced by degradation of glucose during glycolysis salt of pyruvic acid.

pyruvic acid Colorless organic liquid formed as an intermediate in several metabolic pathways (e.g., fermentation and end product in glycolysis).

quorum sensing Social phenomenon known best in populations of microbial organisms that involves synthesis and secretion of signal compounds that induce physiological or behavioral changes, e.g., induction of light in minimal populations of luminescent bacteria.

receptor 1. Specialized cell or group of nerve endings that responds to sensory stimuli. 2. Molecular structure or site on a cell that binds with chemicals (e.g., hormones, antigens, drugs, neurotransmitters) to mediate specific responses.

redox gradient In biogeochemistry, stratification of various oxidizing and reducing agents according to their redox potential over a given area or depth.

replicon DNA polymer capable of self-replication. Large replicons include genophores and chromosomes; small replicons include plasmids, viruses, transposones, and episomes.

res cogitans Thinking thing; in Descartes' dualism, a characterization of the mind or soul as a thing whose sole function is to think.

res extensa Extended thing; Descartes' characterization of physical existence.

reticular formation Brain structure associated with the wake/sleep cycle and selective sensory stimuli.

retinal Aldehyde of vitamin A (retinol), having a double-bonded oxygen in place of an OH alcohol group. After undergoing a conformational change by reaction with light, retinal can bind to photosensitive proteins called opsins, which are involved in various organismal responses to light such as phototaxis, animal vision, and some light-sensitive bacterial metabolic pathways.

rhizobia Microscopic bacteria found in (and inducing the formation of) nodules on the roots of legumes. Rhizobia form a symbiotic relationship with legumes by fixing nitrogen into usable form for the host plant and feeding off its photosynthate.

rhizoplast Striated proteinaceous structure that extends from the kinetosome/centriole to the nucleus (or cytoplasmic microtubule-organizing centers in various protoctists). The nuclear connector portion of the karyomastigont.

ribonucleotide derivatives Compounds metabolized or modified from ribonu-cleotide bases.

riboswitches Parts of messenger RNA molecule that can influence gene expression by binding to small target molecules. They are involved in metabolite-mediated transcription control.

ribozymes RNA molecules that act as enzymes; they catalyze metabolic reactions in the absence of proteins.

RNA Ribonucleic acid, macromolecular component of all life on Earth. A long-chain nucleotide polymer, usually single-stranded, it is essential for translation, the synthesis of proteins from the coded gene sequence.

rotifers Members of the Phylum Rotifera in Kingdom Animalia microscopic, primarily fresh water aquatic animals that bear a wheel-shaped crown of project-ing cilia used for movement and ingestion.

ROY colors Red, orange, and yellow.

sensory transduction Conversion of a stimulus from one energy to another, usually from mechanical or electromagnetic to chemical, e.g., the conversion of light in a photovisual pathway to a biochemical signal in the form of ATP.

sporulation Sporogenesis; the differentiation of spores, e.g., endospores in *Clostridium* and *Bacillus* or the production of spores as products of meiosis in plants.

statolith Intracellular organelle associated with gravity perception (e.g., in plants).

strepsirhini Kingdom Animalia, Class Mammalia, and Order Primates, members of a primate suborder that includes tooth-combed Madagascar inhabitants, e.g., lemurs, lorises, and bushbabies.

stromatolites Laminated or domed sedimentary rocks generally composed of calcium carbonate formed by microbial communities that trapped, bound, and precipitated sediment in an orderly fashion. Stromatolites are microbialites important for reconstruction of the history of early life on Earth.

sulfidogenesis (or methanogenesis) Production and release of sulfide (H_2S) or methane (CH_4), respectively, by microbial metabolism.

superorganism Social colony or well-integrated community that functions as a single organism in the flow of energy, food, and information among its members.

suprachiasmatic nucleus (SCN) Brain structure in vertebrates that generates circadian rhythms by gene product/negative feedback loops in specialized cells, which then relay information to the rest of the animal. The SCN controls rhyth-mic release of a small organic compound, melatonin, that via the bloodstream carries the signal to the body.

symbiogenesis Permanent or cyclical physical association of life forms (bionts) that results in the evolutionary formation of new behaviors, biochemicals, cell organelles, new organs, or new species.

symbiotrophy Nutritional mode in which a heterotrophic symbiont exchanges organic nutrients with a biont of another species.

synodic month   Lunar month; period between successive new moons (29.531 days).

syntrophy   Production of a metabolic product by one biont that serves as the substrate for the second biont (e.g., degradation of a toxic compound to release otherwise unavailable nutrient).

tapetum   Layer of reflective cells behind the retina that amplify incident light in mammalian nocturnal vision.

telencephalon   Brain structure in mammals that is involved in movement, olfaction, learning, and communication.

teleology   Concept of goals and ultimate purpose (e.g., the idea that organisms are designed toward a predetermined destiny).

temporal lobe   Brain structure in the cerebral cortex region involved in processing of auditory or visual information and memory formation.

tensional integrity (tensegrity)   Structural integrity based on a synergism of balanced tension and compression mechanical components.

thermodynamics   Physical science of energy flow, heat, and transformation. Its first law—the law of conservation of energy—states that energy changes but it is not created or destroyed. Its second law states that the quality and usefulness of energy diminishes as heat irreversibly dissipates into the cool.

thiazolium ring   Organic compound that contains sulfur and nitrogen, a component of the B vitamin thiamine.

thioesters   Any of several classes of organic compound in which one or two oxygen atoms of the ester group are replaced by sulfur atoms.

topoisomerase   Enzyme that facilitates DNA replication by catalyzing DNA conformational changes by winding and unwinding the helix.

tornaria   Kingdom Animalia, Phylum Hemichordata; the free-swimming, ciliated, marine larval stage in the life history of the class Enterpneusta (acorn worms).

totipotent   1. Ancestral organism capable of giving rise to new species by direct filiation. 2. Stem cell, e.g., a cell that is able to divide and differentiate into all the specialized cells in an adult animal or plant.

transdisciplinary   Beyond the confines of any academic discipline.

transducer   Device or process that converts one form of energy into another—e.g., gravitational-kinetic energy of a waterfall into generation of electricity in hydroelectric plants or light into chemical energy in the phosphorylation of ATP.

transduction   1. Transfer of small replicated DNA from one organelle or cell to another, typically mediated by small replicons. 2. Conversion of one form of energy to another.

trimer   In chemistry or material science, a polymer or polymeric molecule, consisting of three similar monomers.

Trochophore   Kingdom Animalia; the free swimming, larval stage in the life histories of animals found in more than three phyla (e.g., Annelida, Mollusca, Bryozoa) that has circlets of cilia.

unidirectional   Moving or allowing movement in one direction only, as in genetic transfer in prokaryotes; genes pass from donor to recipient bacterium only in the conjugation act.

viscometer (viscosimeter)   Instrument that measures the viscosity of a fluid.

Wernicke's area   Brain structure of the upper left temporal cerebral cortex of mammals; it is involved in simple verbal and intraspecific non-verbal communication.

Xanthophyte   Kingdom Protoctista, Phylum Xanthophyta; the yellow-green algae in the subkingdom Heterokonta (Stramenophiles).

About the Authors

Celeste A. Asikainen is the administrator of the Margulis Laboratory and a doctoral student in the Department of Geosciences at the University of Massachusetts at Amherst. Her work has been published in the *Journal of Paleolimnology* and in *Proceedings of the National Academy of Sciences*.

Arturo Becerra is a professor in the College of Sciences (Facultad de Ciencias) at the National Autonomous University of Mexico (UNAM). He teaches courses in evolutionary biology and molecular evolution at UNAM.

Nathan Currier is a classical composer who has been recognized with numerous honors, including the Rome Prize and an Academy Award for lifetime achievement in music from the American Academy of Arts and Letters. He resides in Virginia, where he serves on the faculty of the University of Virginia at Charlottesville.

William Day, an Indiana native, holds a Ph.D. in chemistry from McGill University. He is fascinated by the origins of life problem.

Luis Delaye investigates the early evolution of mechanisms against oxygen damage in microorganisms as a postdoctoral student at the Cavanilles Institute for Biodiversity and Evolutionary Biology.

Eshel Ben-Jacob is a world-renowned physicist who has made outstanding contributions to microbiology and neurobiology by applying mathematics and physical principles to living phenomena. He holds the Alex Maguy-Glass Professor of Physics of Complex Systems Chair at Tel Aviv University.

Victor Fet is a professor in the Department of Biological Sciences at Marshall University in Huntington, West Virginia. He immigrated to the United States from Russia in 1988. He co-authored the definitive work on scorpion systematics and co-edited monographs on the biogeography and ecology of Turkmenistan and Bulgaria. He has published three books of poetry in Russian.

John Hall holds a research associate professorship in the Department of Geosciences at the University of Massachusetts at Amherst.

Robin Kolnicki teaches small laboratory and large lecture courses in genetics, human biology, and general biology in the Biology Department at Framingham State College in Massachusetts. She has undertaken field work, concerned primarily with primates, in the Central African Republic, in Sri Lanka, and in

Madagascar. Having deciphered the karyotypic histories of lemurs and equuids, she is researching bat chromosome evolution.

Wolfgang E. Krumbein   is counted among the founders of geomicrobiology and biogeochemistry, new scientific fields that are especially relevant to global climate and planetary biology. He has published more than 400 articles and book chapters, with many collaborators. A founding member of the Institute for Chemistry and Biology of the Marine Environment and the Institute of Philosophy, both at the University of Oldenburg, he recently published a textbook on fossil and extant biofilms.

Laurie Lassiter   has presented more than a dozen papers at the Bowen Center for the Study of the Family in Washington. She maintains a private practice and coaches parents in principles of Bowen theory at the Child Guidance Clinic in Springfield, Massachusetts. For the past decade, she has studied the social and community life of microorganisms with Lynn Margulis at the University of Massachusetts at Amherst.

Antonio Lazcano   is a professor at the College of Sciences (Facultad de Ciencias) at the National Autonomous University of Mexico (UNAM) in Mexico City. He has published several books in Spanish, including *The Origin of Life*, which sold more than 600,000 copies. Lazcano is first Latin American scientist to serve as the president of the International Society for the Study of the Origin of Life.

Paul Lowman   is a geologist at NASA's Goddard Space Flight Center in Greenbelt, Maryland. Hired in 1959, Lowman was the first geologist at Goddard. He was principal investigator for several terrain photography experiments on the Mercury, Gemini, and Apollo Earth orbital missions. He took part in the analysis of lunar samples returned by the Apollo Moon missions, and the lunar remote sensing experiments of Apollo missions 15 and 16. He is the author of *Exploring Space, Exploring Earth: New Understanding of the Earth from Space Research.*

James MacAllister   is a graduate student in geosciences at the University of Massachusetts at Amherst. He spent three decades engaged in all aspects of technical and artistic video production. As a freelance producer and consultant, he made documentaries for the Public Broadcasting Service and earned honors from the Health Science Communications Association, the Society for Technical Communication, and the New England Chapter of the American Medical Writers Association.

Andrew Maniotis   is a research professor in the Department of Pathology at the University of Illinois at Chicago. Currently, he and his colleagues test the reconstitution and dynamics of microsurgically isolated genomes from cells derived from both normal tissues and from malignant human tumors in an attempt to discern fundamental differences among different types of genomes that are responsible for "vasculogenic mimicry," a phenomenon, discovered in his lab, whereby aggressive tumors construct their own perfusion channels independently of new blood vessel formation associated with tumor growth.

Lynn Margulis,   distinguished university professor in the Department of Geosciences at the University of Massachusetts at Amherst, received the 1999 National Medal of Science from President William Clinton. She has been a member of the

United States National Academy of Sciences since 1983 and of the Russian Academy of Natural Sciences since 1997. With students and colleagues, she has written more than 50 books and chapters for scientists, students, and the general public. Margulis is known for her contributions to Gaia theory, James Lovelock's proposal that interactions among live beings, rocks, soil, air, and water have created a vast, self-regulating system at Earth's surface.

Judith Masters is an evolutionary biologist with a long-standing special interest in primates. A professor of zoology at the University of Fort Hare in Alice, South Africa, she holds adjunct positions at the University of KwaZulu-Natal and at Stony Brook University.

Margaret McFall-Ngai was recruited for a faculty position by the University of Hawaii's Kewalo Marine Laboratory in 1996. By 2004 she had become a professor in medical microbiology and immunology at the University of Wisconsin's School of Medicine and Public Health. Her research interests include symbiotic associations of marine animals with microbes, especially bacteria. In 2009 she was awarded a John Simon Guggenheim fellowship for research in organismic biology and ecology.

Kenneth H. Nealson, a professor of Earth sciences and biological sciences at the University of Southern California, holds the Wrigley Chair in Environmental Studies. He has pioneered the field of modern geobiology, which investigates processes of physics, chemistry, and life as they interact with minerals and metals. At NASA's Jet Propulsion Laboratory in Pasadena, he directs efforts to develop criteria for evidence of extant and extinct life in remote places on Earth and elsewhere in the solar system.

Luis Rico, an artist and a culture activist, specializes in metadisciplinary projects that unite artistic, scientific, technological, and social practices. He served as co-director of the MediaLabMadrid Program at the Conde Duque Cultural Center in Madrid. As the director of E-biolab-Madrid, he works to develop interactive educational and public museum projects that use new information and communications technologies.

Gerhard Roth, a professor of behavioral physiology in the Department of Biology/Chemistry at the University of Bremen since 1976, has been the director of that university's Institute for Brain Research since 1989. The founding Rector of the Hanse-Wissenschaftskolleg Institute for Advanced Study and the president of the German National Academic Foundation, he has published approximately 200 works, including six books.

Frank Ryan is a consultant physician and an honorary Research Fellow in the Department of Evolutionary Biology at Sheffield University. His book *The Forgotten Plague* was a *New York Times* Book of the Year. Another of his books, *Darwin's Blind Spot*, was an Amazon Featured Book. He has brought the roles of viruses in healthy symbioses to the attention of the scientific community.

Dorion Sagan has been general partner of Sciencewriters for more than 20 years. His recent books, with Eric Schneider, include *Purpose of Life* and *Into the Cool: Energy Flow, Thermodynamics, and Life.* He is also the author of *Death/Sex* (with Tyler Volk) and a co-author of *Up from Dragons: Evolution* *of Human Intelligence* (with John Skoyles) and *What Is Life?* (with Lynn Margulis). He has also published many articles.

Bruce Scofield   is working on a book about astronomy/astrology—about life in its geocosmic context. As the sole owner and employee of a small publishing enterprise, he has published seven hiking guides for the northeast of the United States. He also has authored books on ancient Mesoamerican astronomy/ astrology. He teaches cosmic evolution and Earth history at the University of Massachusetts at Amherst and online courses for Kepler College.

Yoash Shapira   has served in many government positions, most recently as a military attaché for research and development in Washington and as the director of the research and development division of the Israel Atomic Energy Commission. Since 2001, he has been a visiting scientist at Tel Aviv University's Center for Nanoscience and Nanotechnology.

John Skoyles   holds research fellowships at the Centre for Mathematics and Physics in the Life Sciences and Experimental Biology at University College London and at the Centre for Philosophy of Natural and Social Sciences at the London School of Economics. Among the concerns of his scientific publications are autism, human brain size, the replicative nature of speech units, reading, and the origins of human bipedality and of speech. He is a co-author, with Dorion Sagan, of *Up from Dragons: Evolution of Human Intelligence.*

Robert Sternberg,   a documentary filmmaker and teacher, directs the master's program in Science Media Production at Imperial College London, where he prepares scientists for careers in broadcast media. His film *Hopeful Monsters* chronicles Donald Williamson's work on the chimeric evolution of marine animals and their larval stages.

Alfred I. Tauber   is the Zoltan Kohn Professor of Medicine, a professor of philosophy, and the former director of the Center for Philosophy and History of Science at Boston University. He is the author of *The Immune Self: Theory or Metaphor?*, *Confessions of a Medicine Man: An Essay in Popular Philosophy*, *Henry David Thoreau and the Moral Agency of Knowing*, and *Patient Autonomy and the Ethics of Responsibility*, and a co-author of *Metchnikoff and the Origins of Immunology* and *The Generation of Diversity: Clonal Selection Theory and the Rise of Molecular Immunology.*

Sonya Vickers,   a retired teacher of biology and ecology, lives her dream of lifelong learning. She leads field trips worldwide. With special interest in freshwater zoology, she has given seminars on microscopy and has introduced many to the subvisible universe of life. Her co-authorship has been crucial to the communication of Donald Williamson's ideas on genetic mergers in animal evolution.

Peter Warshall   owns and runs a consulting firm and has worked for the United Nations, for the United States Agency for International Development, for various business firms, for non-governmental organizations, and for local communities. Currently serving as a biologist for the Northern Jaguar Project (www.northernjaguarproject.org) and as co-director of Dreaming New Mexico Project, he was editor-in-chief of *Whole Earth Magazine* for 25 years.

Jessica Hope Whiteside, assistant professor in the Geological Science Department at Brown University, focuses on ecosystem evolution, interpreting past environments through field, laboratory, and library research. Her interests range from biological communities to extrinsic forcing of climatic control driven by environmental catastrophe. She also studies the Eocene sedimentary and igneous rock record of the Western United States.

Donald I. Williamson specialized in the study of marine plankton, especially the animal larvae, at the University of Liverpool's Port Erin Marine Laboratory. He served as Senior Specialist in Decapod Crustacean Larvae for collections for the International Indian Ocean Expedition. He is the author of two books, *Larvae and Evolution* and *The Origins of Larvae* and of many papers.

Dianne Bilyak, our developmental editor, has had poems, interviews, and stories published in *Memoir(and)*, in the *Massachusetts Review*, in *Drunken Boat*, in *Meat for Tea*, and in the *Tampa Review*. A co-founder of the Arts and Literature Laboratory in New Haven, she holds a master's degree from Yale Divinity School. She has been poetry editor of the literary journal *Peregrine* since 2008. Her book of poems *The Length of the Net* has been accepted for publication.

Index

Adaptation, 11, 13, 17, 20, 31, 61, 136, 156, 157, 201, 210, 204, 215, 252, 253
Adenosine triphosphate (ATP), 23, 27, 28, 31, 141, 142, 292
Adenylyl cyclase, 36–42
Alarmones, 5, 35–43, 48, 52, 168, 291
Algae, 6, 7, 93, 96, 101–104, 111, 114, 115, 124, 127, 144
Altruism, 72, 87, 88, 103
Amino acids, 10, 23, 24, 28–32, 36, 39, 57, 113, 140, 291
Animals
  aesthetic zeal, 148, 149
  behaviors, 8, 10, 46, 65, 116, 181, 215, 216, 227, 231, 237–239, 257, 265
  biological clocks, 115, 116
  choice-making, 6, 8, 245, 247
  chromosome evolution, 7, 8
  consciousness, 3, 13, 221–231, 238, 241–243, 247, 250
  eyes, 139, 141, 146
  fission events, 178–181
  grooming, 239, 252, 253, 255
  groups, 281–287
  group selection, 72, 85, 86, 87, 89
  immune system, 199–205
  in food web, 146–148
  microbial partnerships with, 19, 20, 47, 96, 100, 159, 166, 199–201, 211, 215
  purposeful, 94, 245–249

  sensation, 9, 161, 164–166
  social networks, 251–257
  taboo, 239
  tools, 228, 229
  triangle hypothesis, 76–78, 82–87
Archaebacteria, 4, 21, 45, 104, 111, 114, 414, 159, 163, 164, 166, 209, 210, 246, 291
Archaeprotists, 166, 292
Archean eon, 23, 27, 33, 38, 62, 70, 93, 99, 103, 127, 139–141, 259, 292
Aristotle, 110, 249
Art, 259–266
Atmosphere, 11, 23, 55, 92, 95, 99, 114, 117, 118, 124, 126, 127, 130, 134–140, 144, 210, 263, 264
Awareness, 1–3, 8, 9, 13, 77, 82–85, 144, 149, 221, 222, 227, 228, 241, 242, 247, 261–264

Bacteria
  altruism, 87–90
  autotrophy, 105, 143, 202
  behavior, 1, 3, 10–13, 45, 46, 49, 55–58, 61, 68
  calcifying, 65–68
  cells, 1–3, 30–33
  circadian rhythms, 113, 114
  cognition, 9, 10, 55–62
  collective memory, 56
  colonies, 55–63, 75, 87, 89, 162, 173
  cooperative behavior, 55–58

Bacteria (cont.)
 in evolution, 1–4, 8, 119, 209, 210
 groups, 281, 282
 group capabilities, 5–7
 and immune system, 199–207, 217
 insects and, 19, 20, 186, 202, 203,
  212
 intestinal, 105, 210–214, 217
 information-processing systems,
  55–58, 240
 learning, 61, 62
 methylation-demethylation, 49
 non-living origins, 5
 oxygen-respiring, 99, 104, 209,
  210
 photosynthetic, 50, 99, 143
 selection, 72, 85, 86, 87, 89, 98,
  103, 104
 self-engineered organization, 58–61
 sensibilities, 3, 35, 45–52, 55–57,
  119, 153, 161–165
 social, 5, 6, 56, 58, 87–89
 structure-building, 6, 57–60, 63–65,
  68–70
 symbiotic, 19, 20, 99, 143, 163,
  165, 186, 211, 217
 thermodynamic engines, 55–58
Behaviors, 3, 10, 13, 45, 57, 65, 68,
  71–78, 81–89, 106, 116, 248
Biological rhythms, 7, 109–121
Bioluminescence, 51, 52, 114, 138,
  139, 292
Biosphere, 9–11, 42, 92, 94, 99, 100,
  105, 135–148, 217, 263, 265,
  292
Bohm, David, 243, 263
Bonding relationships, 237–239,
  253–257
Bowen, Murray, 6, 71–90
Brain
 chimerization, 233–240
 cognitive abilities and, 1, 221,
  227–231
 human, 8, 9, 221–231, 234
 plasticity, 233–237
 reptile, 229, 230
 sensation, 221, 226–229

Cambrian period, 103, 130, 132,
  139, 145, 293
Centromere, 173–181, 293
Centrosomes, 21, 157, 161–164, 293
Capitulation, 76–78, 82
Carotenoids, 132, 133, 146, 147, 293
Castells, M., 259, 265
Cell membranes, 10, 12, 13, 27, 141,
  160
Cells
 communication between, 1–3,
  51–62,
 division, 164–170, 175–179, 211
 evolution, 4, 17–21, 26–28, 99, 102,
  169
 motility, 45, 159–166, 211
 nucleated, 4, 7, 42, 45, 46,
  157–166, 211
Centrioles, 161–165, 293
Chemosensitivity, 10, 49, 293
Chemotaxis, 48–51, 60, 294
Chimerical union, 4, 7, 8, 154, 237,
  294
Chlorophyll, 88, 92, 132, 133,
  138–147
Chromosomes
 chromolinkers, 168–172, 294
 discontinuity, 169–172
 evolution, 4, 8, 45, 157, 167–172
 fission, 20, 174–181
 mitotic checkpoint control, 170, 171
 split, 173–181
Ciliates, 10, 102, 111, 114, 115, 162,
  166, 170, 188, 211, 294
Circadian rhythms, 111–117, 120,
  121, 294
Cognition, 8, 9, 55–62, 200, 201,
  205, 221, 222, 228–231
Collective consciousness, 56,
  251–257
Colonial structures, 56–61
Color, 130–134, 144–149, 251–253
Common ancestral stock hypothesis,
  187–190
Communication
 brain chimerization and, 236, 237
 cellular, 9, 10, 51, 52

biochemical, 47, 48, 61, 62
microbial, 5, 6, 55–62
in self-engineered organization,
  58–61
Communities
consciousness, 51, 52, 251–257
formation, 9, 55, 57, 161–166
nested, 91–106
structures, 63–70
Competition, 95, 99, 100, 104
Convergence, 194–196
Cosmic cycles, 7, 109–121
Cretaceous period, 147, 212, 294
Culture
brain chimerization and, 238–240
networks, 259–266
Cyanobacteria. *See* Bacteria
Cybernetics, 56–58, 93, 213
Cyclic adenosine monophosphate
  (cAMP), 35–43, 86, 87, 294
Cytoplasm, 13, 49, 112, 115, 157,
  209, 294

Dadaism, 263, 264
Daisyworld model, 94, 95
Darwin, Charles Robert, 72, 75, 84,
  87, 96–98, 102, 104, 105, 110,
  153, 156, 157, 161, 162, 183–185,
  217, 261
Darwinism, 96, 97, 153, 156, 230,
  235
de Bary, Anton, 95, 96, 161
Descartes, René, 242, 243, 244, 245
Diatoms, 63–68, 114, 295
Dinomastigotes, 114, 143, 170, 295
Disease, 5, 19, 165, 208–216
Divergence, 174, 179–181, 192
DNA, 7, 8, 10, 17, 26, 31, 36, 43,
  98, 115, 142, 164, 167–180, 190,
  195, 211, 247

Earth
geomagnetic field, 118–120
as open thermodynamic system, 93,
  94, 208
photobiological history, 130,
  139–149

rotation and rhythms, 109, 111,
  118, 119, 120
tectonic plates, 7, 95, 123–127, 173
Ecosystems, 11, 99, 100, 105, 208,
  248, 258, 265
Electron transfer, 26–28, 32, 131
Empathy, 221, 228, 231, 241, 242
Endosymbiosis, 143, 217
Energy flow
consciousness and, 245–250
metabolism-mediated, 167
in origin of living systems, 23–33,
  96, 105
thermodynamics, 212, 214
Energy gradients, 93, 245–250
Energy-growth core, 25–31, 295
Energy taxis, 49, 50
Entropy, 55–58, 208–210, 217, 218,
  245, 246
Environments
adaptation to, 99, 100, 195, 215,
  228, 231
cosmic cyclal and rhythms, 7, 104,
  109–121,
Earth types, 3, 4, 6, 31, 65, 129
fitness with, 13, 105, 214, 235, 264
health,
life alteration, 5, 11, 13, 31, 32, 35,
  36, 55, 61, 62, 68, 101
sensibilities, 1–3, 6, 11, 45–52,
  56–58, 72, 79, 85–89, 131, 132,
  139–147, 153, 161, 167, 172,
  199–203, 234, 248,
Eubacteria, 4, 36, 45, 47, 104, 114,
  159, 166, 209, 246, 281, 282, 296
Eukaryotes
circadian rhythms, 111, 112, 115,
  116
evolution, 99, 100, 104, 142, 163,
  165, 166, 186, 199
groups, 282–287
movement, 159, 161
and prokaryotes, 46, 47, 157, 158
Evolution
of alarmones, 38–43
brain-symbol chimeras, 237, 238
eyes, 7, 9, 129, 130, 161

Evolution (cont.)
fission and, 173–181
interspecies hybrids, 183–197
lag, 215
of life without DNA, 36–38
light sensing, 129–149
source, 3, 95
viruses, 17–21
Evolutionary novelty, 3, 7, 9, 145,
  188, 231, 261

Ferredoxins, 27, 28
Ferromanganese nodules, 68–70
Filiation, direct, 103, 104
Flagella, 48–50, 246, 296
Flavonoids, 132, 141, 147, 148, 296
Food web, 133, 134, 146–149
Foraminifera, 65, 70, 102, 118, 247,
  296

Gaia, 6, 7, 11–13, 91–96, 99–102,
  105, 106, 124–127, 263–266, 296
Galagos, 251–257
Gametes, 118, 164, 175, 176
Genophoric-genomic system, 8, 20,
  167–172, 297
Geographical barriers, 174
Geological time scale, 97, 290
Geomagnetic field, 118, 119
Geophysical cycles, 109–112, 120,
  121
Gould, Stephen Jay, 97, 103, 185
Greenhouse effect, 126, 127

Hadean eon, 130, 138, 139, 142,
  144, 297
Health, 8, 13, 208, 209, 214–218
Heterocysts, 88, 89, 113, 297
Holobionts, 18–21, 297
Humans
behaviors, 8, 71–90, 243 244, 259
chromosomes, 8
communication, 223,
consciousness, 3, 228–231, 241
distance, 76–78
DNA, 8
emotional system, 72–90

evolution, 199, 200, 233, 236–239,
  265
free will, 243–245
psychology, 6,
social, 92, 106, 259, 266, 263
and symbols, 97, 98, 181–197

Immune systems, 5, 8, 19, 199–207,
  216, 217
Inheritance, 96–98, 157, 169–172
Insects, 115, 116, 120, 138, 145,
  148, 161, 183–196, 202, 203, 210,
  212, 217
Insect-virus associations, 19, 20
Invertebrates, 201–205
Iron-sulfur clusters, 27, 28

Karyomastigont, 21, 162, 166, 166,
  166, 298
Karyotypic fission, 173–181
Kinetochores, 173–181
Kinetosomes, 161–165
Kozo-Polyansky, Boris Mikhailovich,
  7, 94–99, 153–158, 162

Lamarck, Jean Baptiste de, 96–98
Language, 8, 94, 96, 105, 193, 221,
  226–240
Larval transfer hypothesis, 183–197
Last Common Ancestor (LCA), 10,
  36–39, 298
Learning, 17, 61, 62, 224, 228,
  236–240, 261, 266
Lemurs
evolutionary history, 8, 173, 174
karyotypic fission, 178–180
social networks, 251–253
Light
biospheric, 129–136
infrared, 132, 133, 135
prebiotic, 139, 140
sensitivity to, 52, 115, 129–149,
  200, 234–237
solar, 93, 95, 120, 129, 130,
  134–140
ultraviolet, 132, 135–144
wavelength amplitudes, 7, 132, 133

Limbic system, 222–225, 228, 298
Lipid, 1, 12, 13, 23, 24, 27–31, 63, 298
Lithosphere, 95, 123, 124, 126, 263
Lovelock, James E., 91–95, 125–127, 248, 263, 264
Lunar rhythms, 109, 110, 117, 118, 120

Machiavellian intelligence, 231
Magnetotaxis, 118, 119, 164
Mars, 7, 11, 92, 93, 95, 124–126
Mate recognition, 8, 253–257
Mechanosensitivity, 2, 10, 50, 235, 299
Medicine, 5, 207–218
Meiotic sex, 46, 165
Mendel, Gregor, 97, 98, 102, 169, 299
Mereschkovsky, Konstantin Sergeyevich, 7, 96, 104, 153–157, 161, 162
Metabolism
  autotrophy, 9, 105, 143, 202
  chemoautotophy, 9, 11, 41, 65, 66
  chemoorganoheterotroph, 102
  evolution, 24–26, 30–33, 36, 50
  heterotrophic, 11, 41, 63–68, 143, 202
  necrotrophic, 199
  pathways, 36, 39, 42, 46, 92
  photoautotrophic, 9, 11, 41, 65, 66, 68, 200
  in regulation of climate, 124
  thermoautotrophic, 41
Methyl-accepting chemotaxis sensory proteins, 49–51, 299
Microbes, 6, 47, 48, 51, 63, 64, 68, 71, 72, 87, 162, 163, 199–212, 259, 300
Mind-matter split, 242–245
Mitochondria, 21, 30, 100–104, 115, 157, 165, 186, 207–211, 300
Mitosis, 45, 163, 164, 165, 168–171, 211, 300
Monads, 162, 163, 214
Multicellularity, 46, 100, 104, 300

Mutations
  chromosomes, 176, 179, 180
  random, 3, 9, 95, 98, 103, 167, 183, 195

Natural selection
  and evolution, 32, 94–98, 100, 153, 156, 161, 162, 184, 185, 188, 190, 196, 247
  family and, 71, 72, 82, 84
  fission, 176,
  neural, 234, 235
  parasexual, 217, 218
  sensibilities, 138, 148
Neo-Darwinism, 3, 71, 82, 87, 98, 102–104, 183, 194, 276, 300
Neural plasticity, 233–237, 301
Neurons, 8, 200, 221, 228, 230, 233–236, 261, 262, 265
Nucleotides, 24, 29–31, 37, 301

Oceans, 123–127, 139, 142
Ooids, 63–70, 301
Organophosphates, 26, 28, 32

Pangenesis theory, 96, 98, 157
Phanerozoic eon, 126, 144, 146, 147, 149, 161, 185, 266, 302
Pheromones, 61, 167, 208, 302
Photons, 129–131, 135–147
Photoperiodism, 110, 116
Photoreceptors, 112, 115, 116, 129, 132–143, 146, 147
Photosensory systems, 129–149
Photosynthesis, 7, 30, 46, 88, 113–115, 126, 127, 131, 134, 139–145, 303
Phytochrome, 132, 133, 145, 147, 303
Pigments, 131–133, 139–149
Planetology, 7, 11, 92, 93, 95, 125, 126
Plants
  evolution, 4, 17, 97, 103, 104, 199
  flowering, 147, 148
  groups, 281–287
  movement, 159, 160

Plants (cont.)
  photoperiodism, 116
  photosensing, 7, 46, 131, 134,
    145–147
Plate tectonics, 7, 123–127
Prebiotic molecules, 5, 24, 26, 130,
    139–142
Prokaryotes
  and eukaryotes, 17, 18, 70, 157, 158
  evolution, 3, 4, 99–103, 161–165
  groups, 281–282, 304
  photosensitivity, 130, 131, 138,
    142–144
  sensibilities, 35, 45–52
Preteobacteria, 40, 41, 50, 163, 165,
    168, 304
Proterozoic eon, 4, 103, 127, 130,
    142, 143, 146, 161, 162, 163, 165,
    304
Protoctists
  cell division, 164, 165
  circadian rhythms, 114, 115
  communities and, 101–103
  evolution, 3, 4, 18, 99, 130–132,
    161, 162
  groups, 282–284, 304
  photosensory systems, 142–145
  symbiosis and, 19
Punctuated equilibrium, 4, 8, 97,
    185, 305

Quorum sensing, 51, 52, 305

Ramón y Cajal, Santiago, 261, 262,
    265
Recombination, 9, 170, 201,
    263–266
Reproductive isolation, 177–181
RNA world, 25, 35–43, 306

Same stock theory, 184, 185, 190
Sea-floor spreading, 123–125
Sea urchins, 183, 191–193, 197
Self
  awareness, 1, 8, 221, 222
  differentiation, 71–90
  earliest, 1–3, 30, 31

functional, 81, 82
  maintenance, 12, 32
  sensory, 46–48
  of viruses, 17–22
Selfish gene, 91, 98
Selfishness, 81–83, 103
Self-recognition, 222, 229, 231
Self-reflection, 221, 222
Sensation
  alarmones and, 35–43
  awareness and, 1–3
  in bacteria, 55–62
  collective, 13, 56–58
  earliest, 5, 21, 45–52
  movement and, 159–166
  thermodynamics and, 241–250
Serial Endosymbiotic Theory (SET),
    21, 210, 214
Sexual selection, 162, 217, 218, 256,
    257
Signaling systems, 252–254
Social awareness, 5, 6, 261–264
Social networks
  culture and, 259–266
  group selection, 72, 85, 86, 87, 89,
    98, 103, 104
  and natural selection, 71–90
  primate, 231, 251–257
Solar cycles, 109–121
Solar system, 11, 92, 95, 124, 129,
    136
Specialization, 105, 106, 113, 116,
    136, 230
Speciation, 96, 97, 174–180
Spinoza, Baruch, 243–246
Spirit-matter dualism, 242–245
Spirochetes, 7, 9, 159–166, 209, 211,
    216
Stress response, 35, 36, 58, 61,
    84–86, 167, 168, 195, 208
Stromatolites, 64, 65, 306
Sun-Earth coupling, 134–136
Superorganisms, 4, 57, 210–214, 306
Symbiogenesis, 7, 95–102, 153–158,
    162, 166, 217, 306
Symbiosis, 7, 19–21, 95, 96, 153,
    156, 157, 161, 162, 210–218

Teleology, 248–250, 307
Tetrapods, 221, 229, 230
Thermodynamics
 and consciousness, 241–250
 and environment, 55–58
 open systems, 12, 93, 94, 208, 209
 second law, 208, 209, 245–250
Tidal rhythms, 117, 118

Venus, 7, 11, 92, 93, 95, 124, 125,
 126
Vernadsky, Vladimir Ivanovich, 10,
 11, 102, 265
Viruses, 5, 17–21
Vocal communications, 229, 231,
 236–240
Vocal repertoires, 75, 253–255

Wallace, Alfred Russel, 96–98
Wallin, Ivan E., 3, 97, 100–104, 153,
 154
Water
 environments, 6, 64, 115, 117, 119,
 142
 in life, 2, 10, 12, 13, 23, 32, 48, 52,
 55, 165, 209, 217,
 planetary, 7, 11, 28, 93, 95, 99,
 124–127, 134, 137–140, 144

Printed in the United States
by Baker & Taylor Publisher Services